Technology Activity Guide 1

TECHNOLOGY ACTIVITY GUIDE 1

Glenn Listar

Delmar Publishers Inc.®

Cover Photo
Eaton Corporation

Cover Design
Susan Mathews

Design and Production by
Scharff Associates, Ltd.

Delmar Staff
Executive Editor: Mark W. Huth
Associate Editor: Jonathan Plant
Production Editor: Cynthia Haller

For information address Delmar Publishers Inc.,
2 Computer Drive West, Box 15-015,
Albany, New York 12212

Printed in the United States of America
Published simultaneously in Canada
by Nelson Canada,
a division of International Thomson Limited

10 9 8 7 6 5 4 3 2 1

Library of Congress Cataloging in Publication Data

Listar, Glenn.
Technology activity guide 1.

Accompanied by teacher's guide.
1. Technology. 2. Industrial arts. I. Title.
T47.L497 1987 600 86-24319
ISBN 0-8273-2739-0 (pbk.)
0-8273-2738-2 (teacher's guide)

Contents

Preface

Technology Activity Guide 1 is the first of two workbooks designed to provide creative learning activities for the introduction to technology courses. Now that the shift from traditional industrial arts programs has begun throughout the nation, the need for such books is apparent.

Technology education is a broad-based program that draws its subject matter from many different fields and disciplines. One important inheritance from industrial arts is the emphasis given to hands-on learning activities. Students will develop practical skills in their technology education courses and use them to develop interesting projects. They will have the opportunity to create products of technology and learn about the systems and resources involved in all technological processes.

SCOPE OF THE BOOK

Unit I of the book has two objectives: to introduce the world of technology and its major attributes, and to develop some important background skills. Activities on drawing, designing, and lettering can be assigned as needed to prepare students in these important areas.

Unit II discusses the resources of technology and how they are used. Students will learn about the types of resources, where they are located, and how they are all necessary to technology. An activity on tools helps familiarize students with some of the basic tools they'll be using to create products of their own.

Unit III introduces problem-solving through a selection of activities. The general problem-solving skills developed in this unit will help students to think creatively and approach problems in a logical, systematic way.

Unit IV presents systems, the concept basic to technology education. Activities in this unit demonstrate how every technology consists of systems and subsystems. Open-loop and closed-loop systems are shown and explained.

Unit V shows how technology has affected our lives, for better or for worse. Students will work on activities involving the future possibilities of technology.

Acknowledgments

Technology Activity Guide 1 is the result of several years' experience with technology education students at Greece Athena Junior High School in Rochester, NY. These are class-tested activities that have been popular and successful. They are planned to cover the first semester of an introduction to technology course in conjunction with a classroom textbook such as *Technology in Your World* or *Technology and You,* also from Delmar Publishers.

I would like to thank my associates in the Greece Central School District for their assistance, especially Frank Reed. The following reviewers also provided valuable comments and suggestions: Irvin Randall, George Trombetta, Joseph Santopietro, Tony Lazzaro, Charles L. Smith, Henry C. Strada, Timothy Taraska, and Bernard W. Przelski.

UNIT I

Introduction to Technology

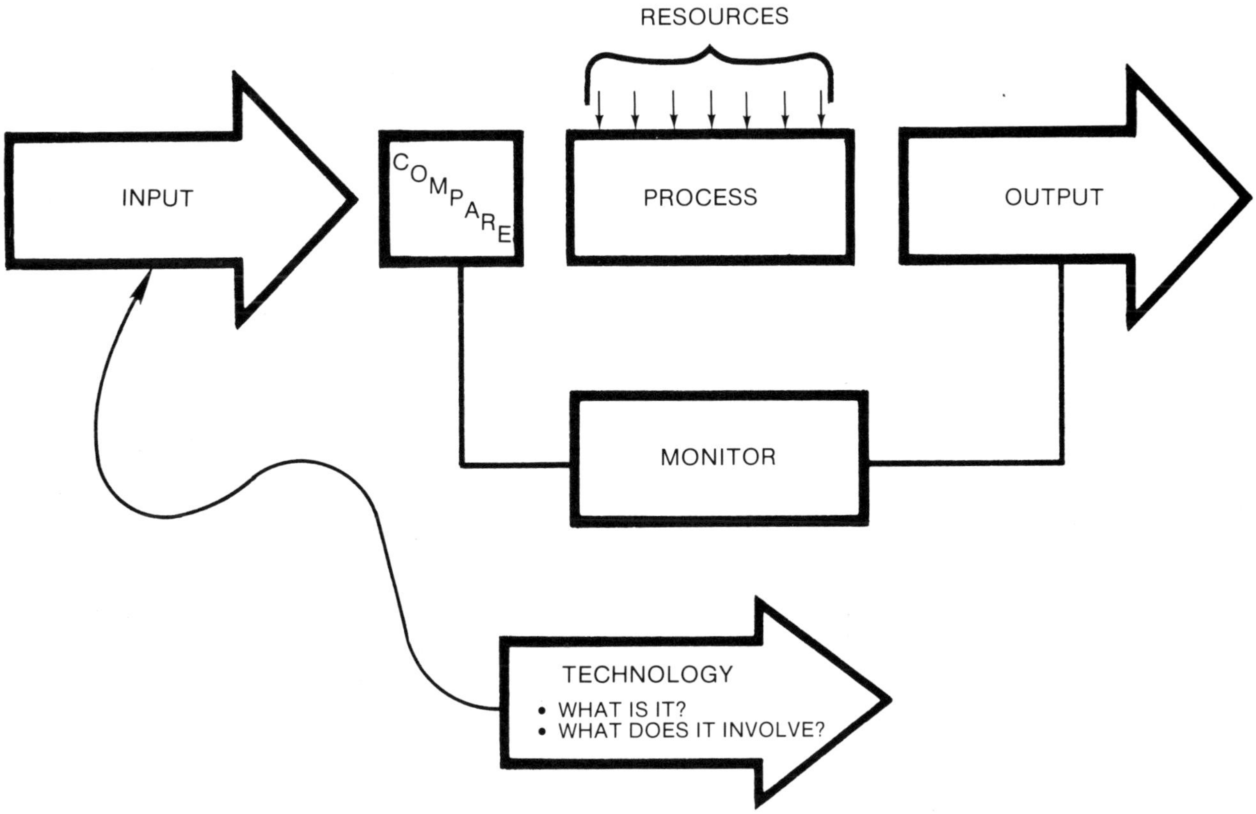

In this unit you will discover that technology is the means by which people change their world. For thousands of years people have developed technologies to make life easier. For instance, they learned to build houses and other shelters that were more comfortable than caves. The wheel was invented and sparked the development of transportation. Refined methods of the production of foods and medicines improved people's health. Better ways to communicate were developed, and the invention of the printing press made it possible to produce books in large numbers. Metals and other materials replaced wood and stone in the manufacturing of tools and weapons. In each of these areas growth in technology slowly took place.

Within the past century, technology has grown more and more rapidly, and our world has changed very much. Now we can construct huge buildings, super highways, and tunnels under the sea. Rockets can transport us into space, and submarines allow us to explore the deepest oceans. Improved fertilizers and pest control have increased the amount of food produced from each acre of farmland. Advanced drugs help us live longer, as well as transplanted organs and implanted artificial body parts. Satellites in space allow us to communicate with distant parts of the world in only seconds. Robots and other new machines manufacture cars and other products safely and efficiently.

How do we learn about such amazing new forms of technology? How can we understand it and put it to use ourselves? The answer is through learning about systems. Throughout this book you will see systems diagrams and hear about the systems approach. It is a way of breaking down complex problems into steps that are easier to work with.

The goal of this unit is to make an Evolution Board or a How Things Work Board. The boards will bring together skills in measuring, drawing, lettering, designing, taking photographs, and gathering information. The activities in this unit will help you learn these important methods. Each phase in making up the boards will be covered in a separate activity.

You can use the chart shown below to plan the technology board your group has decided to make. By planning the phases for your technology board you can use the activities to learn how to do each part successfully.

EVOLUTION BOARD	**HOW THINGS WORK**
Phase 1	**Past**
Phase 2	**Present**
Phase 3	**Future**
Phase 4	**Method to Show Disassembly**

The finished board will be something you've planned and constructed, a product of technology you can be proud of!

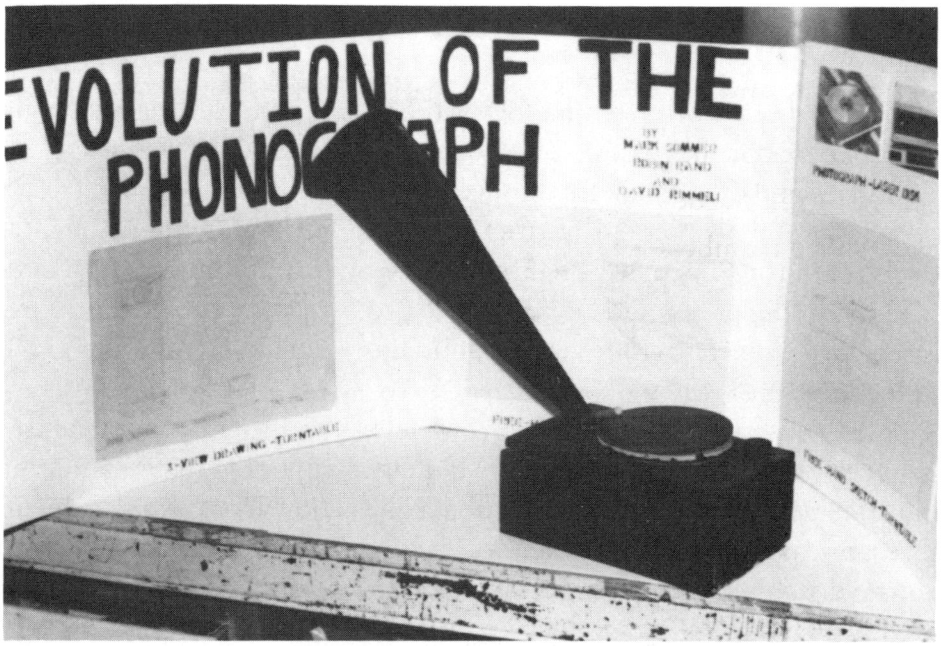

Name	Group Members
Selected Item	

Activity I-1
Planning an Evolution Board

Many items and devices in the world have been modified (changed) since they were first invented. This modification was performed to keep the item up-to-date, improve its performance, remain in the market against competitors, and better meet human needs and wants.

The invention of an item or device from a human need or want and the improvement of that item is an example of technology.

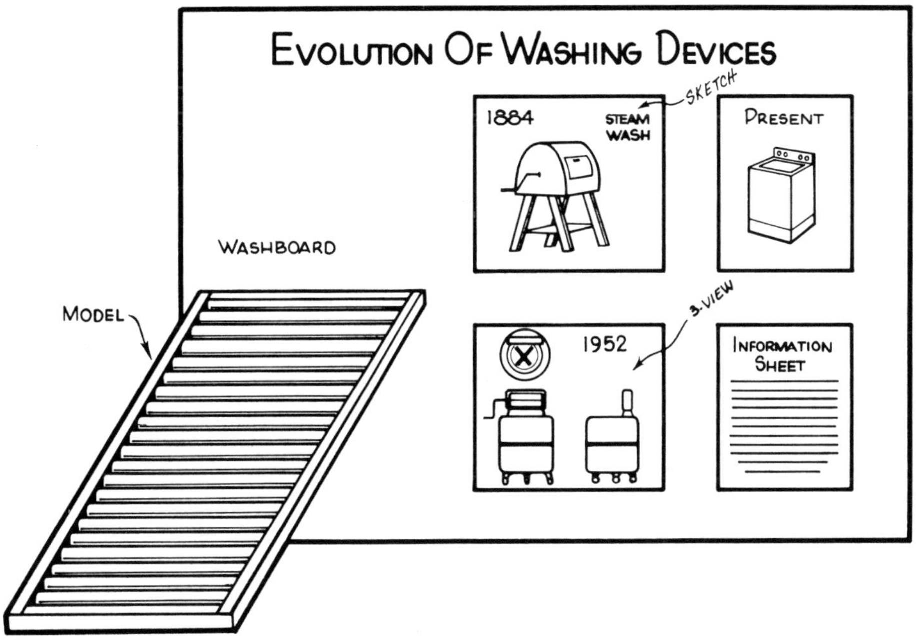

Sample of an Evolution Board

WHAT YOU ARE TO DO:

1. To master the skills needed for this activity, do Activities I-3 through I-16.
2. Determine group members.
3. Select a device, and show at least four aspects of it. Show how the device has changed since it was first invented.
4. You will show each of these changes or aspects by using the activities in this unit. Use a different technology process (activity) to show each aspect. All of these will be displayed on a backboard, such as cardboard. Along with the stages being displayed, you will be writing an information sheet stating who invented the device, the various dates for each of its stages, the type of technology it represents, and the need or want the device meets.
5. Add the various dates to your technology time line and to the computer time line.
6. Present your Evolution Board to the class.

Name	Group Members
Selected Item	

Activity I-2 Planning the How Things Work Board

Many of the items and devices in our world have been changed since they were first invented. The items were invented and produced to meet certain human needs and wants. The first versions of these items may seem crude when compared to their present forms that were modified through the years. These modifications have changed the ways that many of these items operate. But note that the items still perform the same original desired tasks, but the methods in which they do these tasks have changed.

In this activity, you will discover how a specific item works.

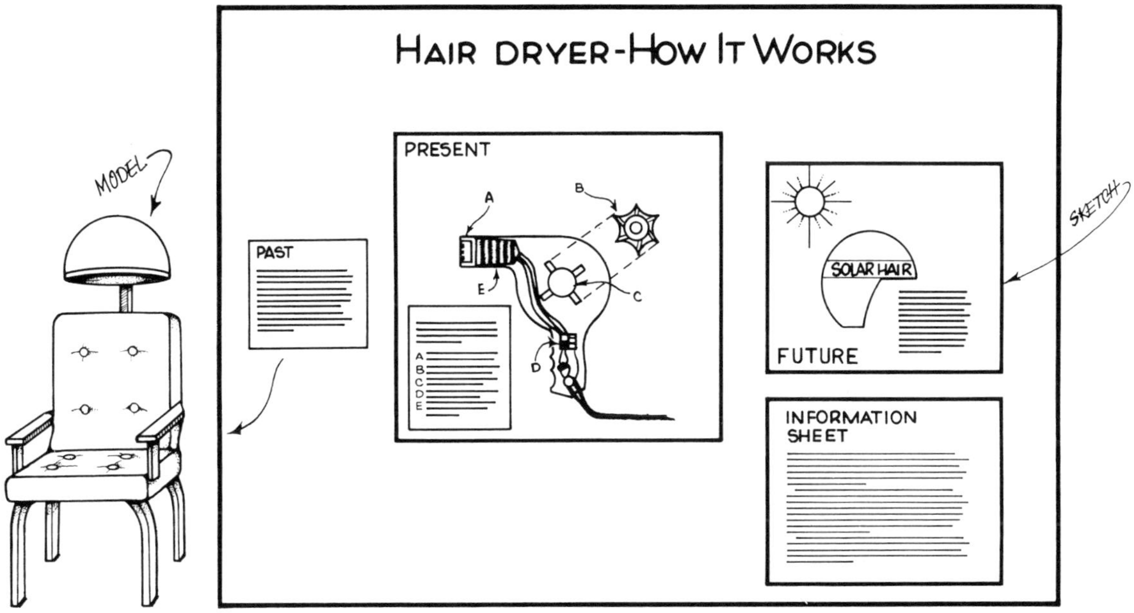

Sample of How Things Work Board

WHAT YOU ARE TO DO:

1. To master the skills needed for this activity, do Activites I-3 through I-16.
2. Determine group members.
3. Select an item (device) that interests you. Find a broken or unwanted item and bring it in to be taken apart.
4. Slowly take the item apart, and photograph or sketch the various steps of dismantling it.

5. Discover what the parts do.
6. The disassembly of the item is to be displayed on a backboard. Using other technology processes, show how the device looked in the past and how it might look in the future.
7. Prepare an information sheet that explains how the item works. Include the dates of the invention (set on time line and in the computer), and the type of technology your device represents.
8. Present your completed How Things Work Board to the class.

Name	Group Members
Selected Item	

Activity I-4
Sketching

The simplest form of design drawing is sketching. Sketches are important b ause th can be made quickly and without any equipment, except a pencil. You should practice etchin til your sketches look neat and you can make them easily.

Pencils come in different hardnesses. Drafting pencils are graded by num ers to how hard they are. Number 6 pencils are hard and make light lines. Number 2 pencils are airl o and make darker lines. A number 2 pencil is good for sketching. Make sure your pencil is arp. g od eraser is also a big help.

When you start, make drawings with drafting tools. It will help if you sk ch he awing first. Professional drafters and designers sketch almost every drawing first to pla it ayo .

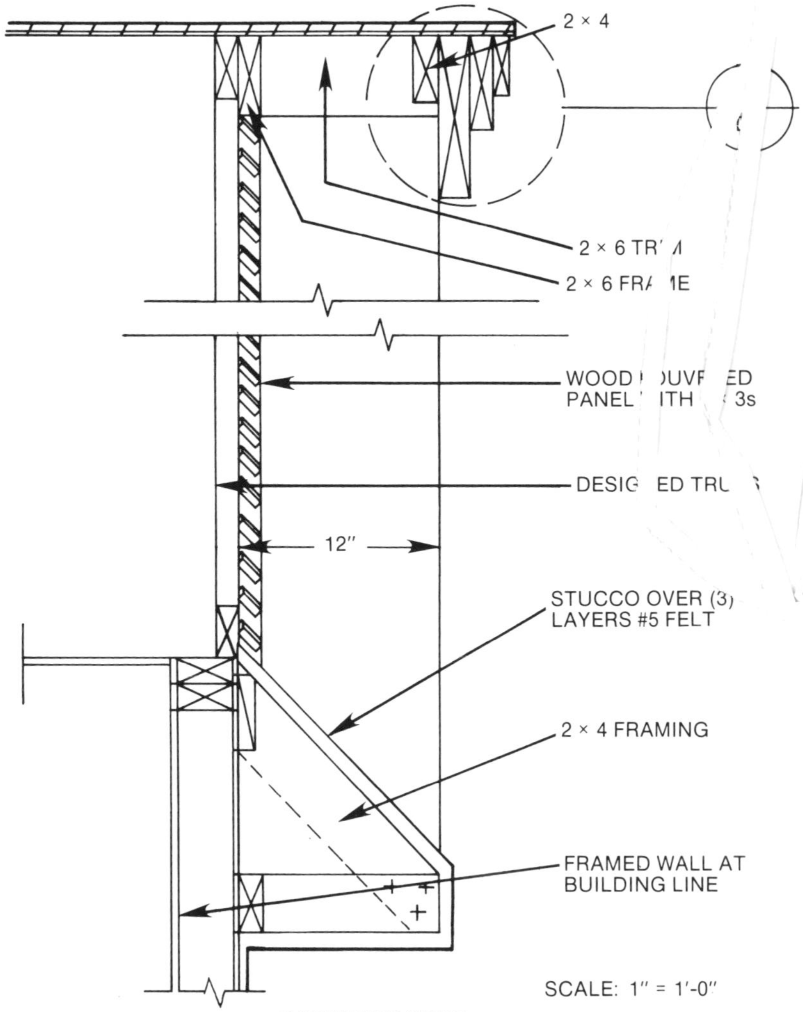

GABLE END VENT

Short straight lines are fairly easy to sketch. To make your longer lines easier, start with a row of dots. Then connect the dots with very light lines. Lastly, darken the light lines.

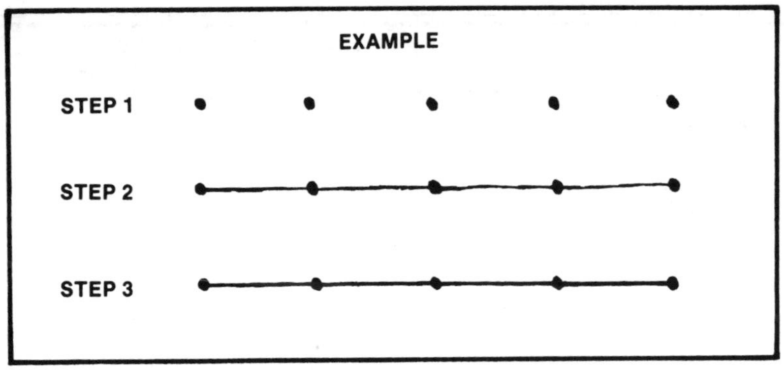

WHAT YOU ARE TO DO:

In the top box below practice connecting the dots and sketching straight lines. In the bottom box sketch 3 more lines to match each of the ones shown.

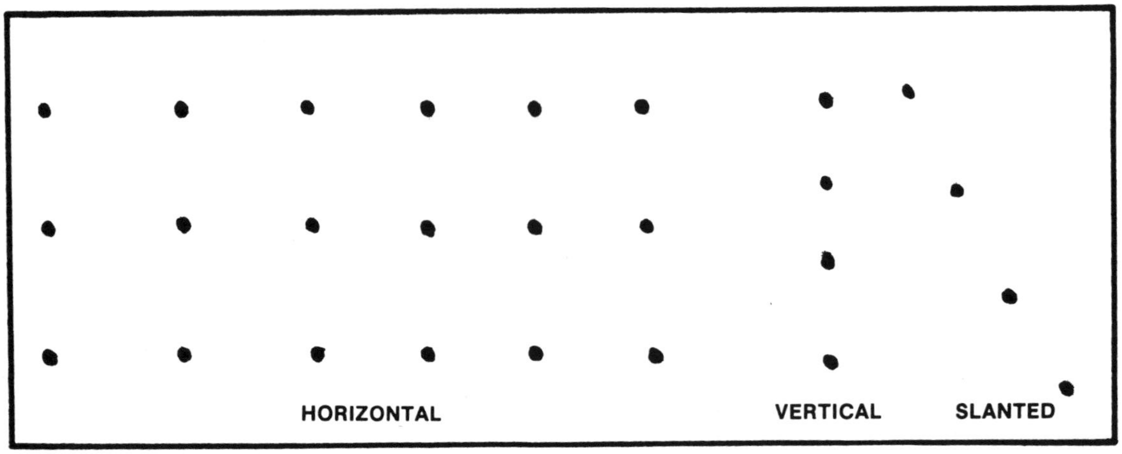

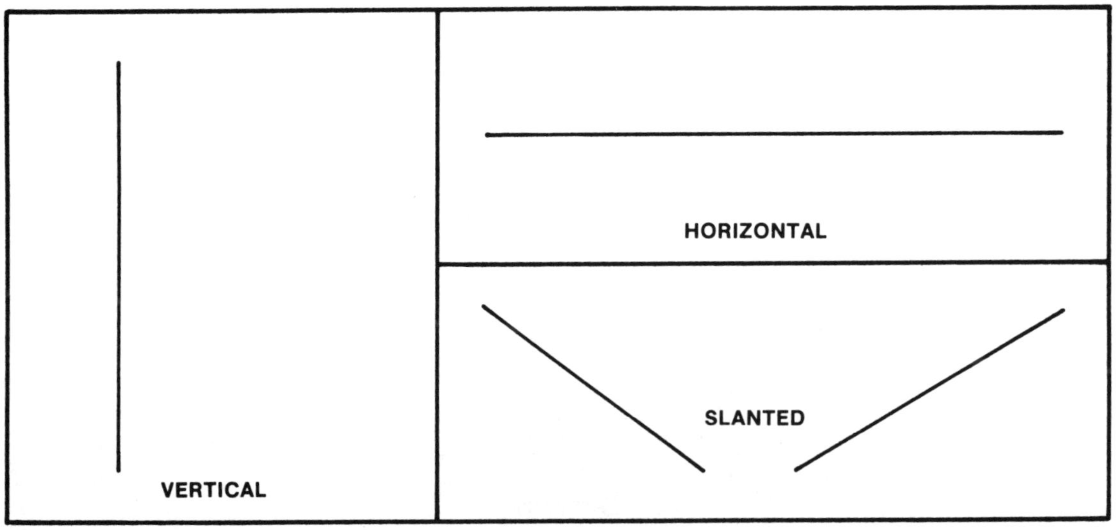

Did you ever notice that if you draw a circle inside a square, it touches each side of the square at its middle? To sketch a circle, start by sketching a very light square. Then mark the middle of each side. Lastly, sketch a curve in each corner of the square. Don't be afraid to turn your paper if it helps.

EXAMPLE

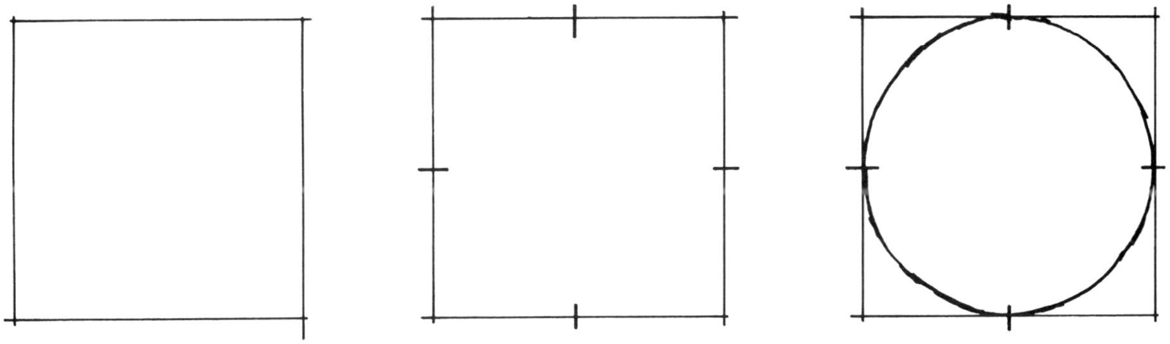

STEP 1 STEP 2 STEP 3

WHAT YOU ARE TO DO:

1. Sketch a circle in each of these squares.

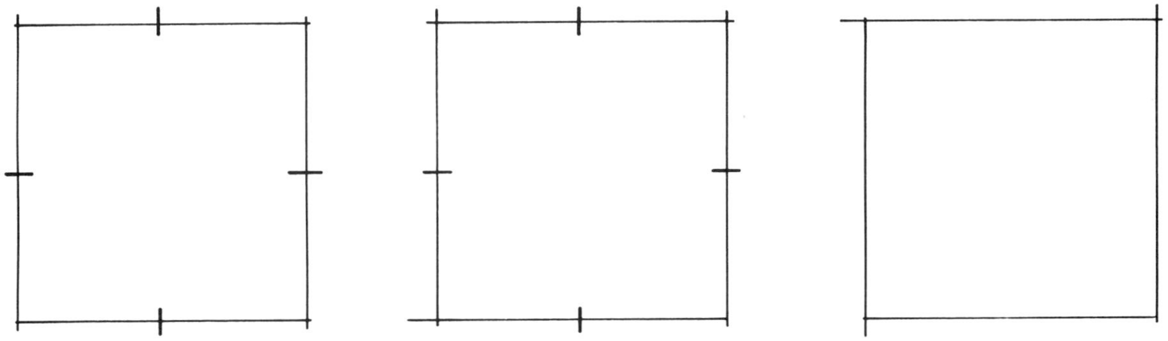

2. Sketch three circles about the same size.

Sketch each of the objects shown, but twice as large.

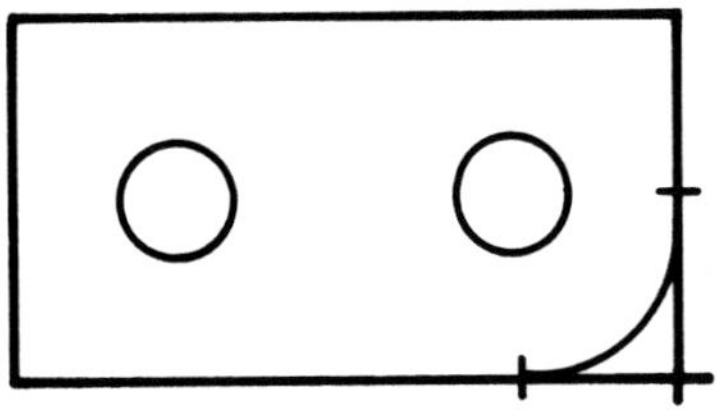

A

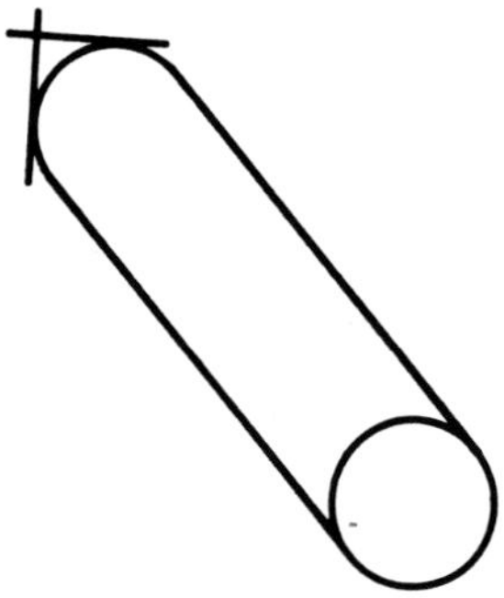

B

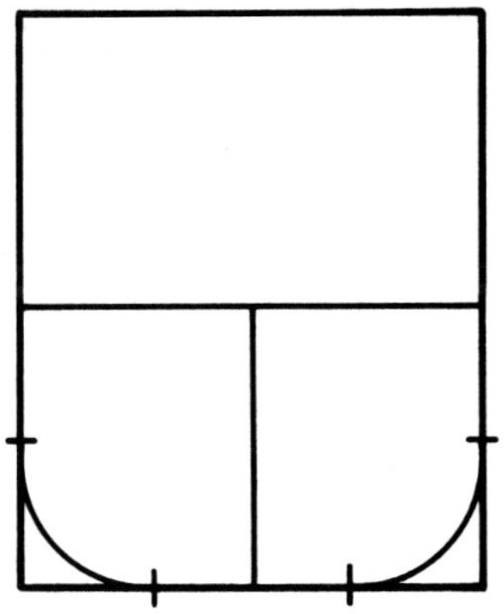

C

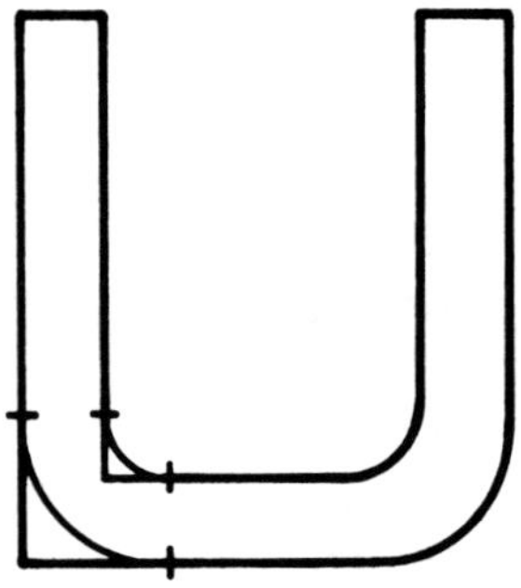

D

Sketching Page

Sketching Page

Name	Group Members
Selected Item	

Activity I-6
Three-View Drawings

Three-view drawings are often called *working drawings.* This is because people build things from them. Working drawings show three separate views of an object, usually the front, top, and the right side.

The *front view* is a drawing of an object looking at it head on. The front view is usually placed in the lower left corner of the drawing.

The *top view* is done looking down on the object and drawing a "bird's eye view" of it. The top view is placed directly above the front view. Projection lines off the front view and gives the length and location of part to the top view. These projection lines are drawn very lightly.

Projection lines off the right side of the top view and meets a 45° line (made by the 45° triangle). Where these lines meet the 45° line, more projection lines are drawn vertically. Once again projections are drawn off the front view, this time to the right. This gives the height of the *right side view.* Projection lines off the 45° line and gives the width in the right side view.

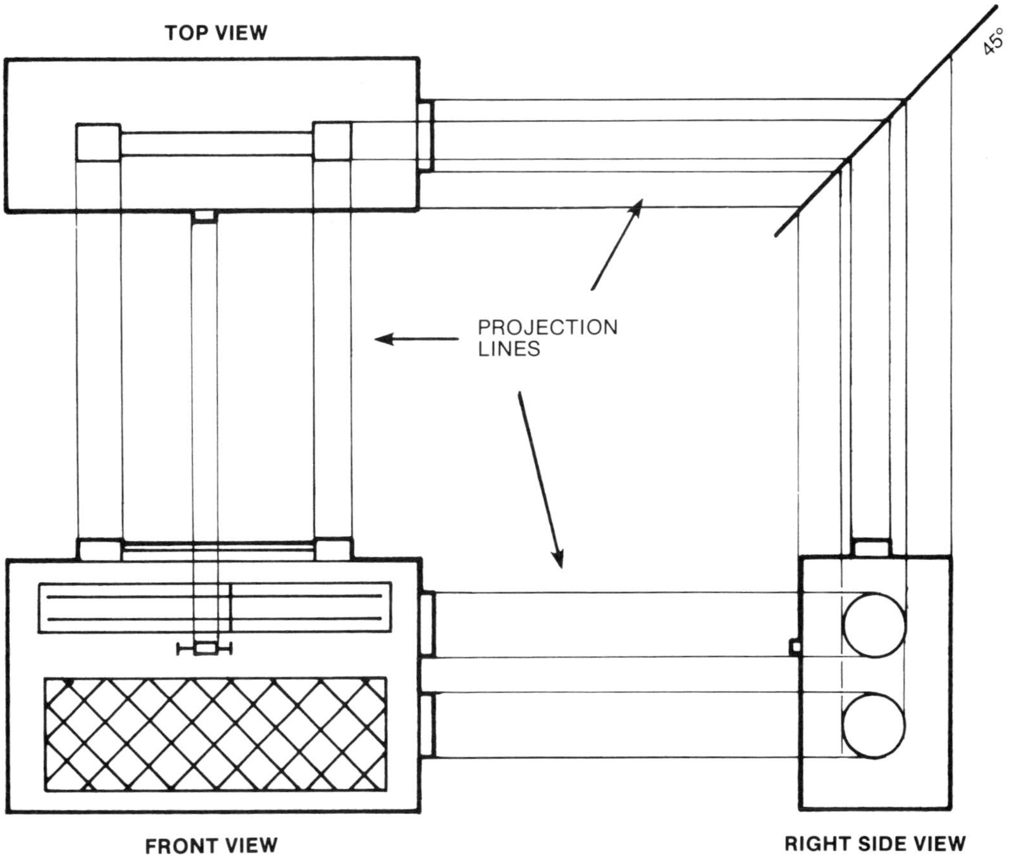

WHAT YOU ARE TO DO:

1. Complete the three-view drawing shown on the next page.
2. Tape the paper on your drafting board.

Note: Many of the items you will be designing will require the three-view drawing. You may wish to use a three-view drawing for your group's Evolution Board or How Things Work Board. Do not try to draw a difficult object at first.

Three-View Drawing

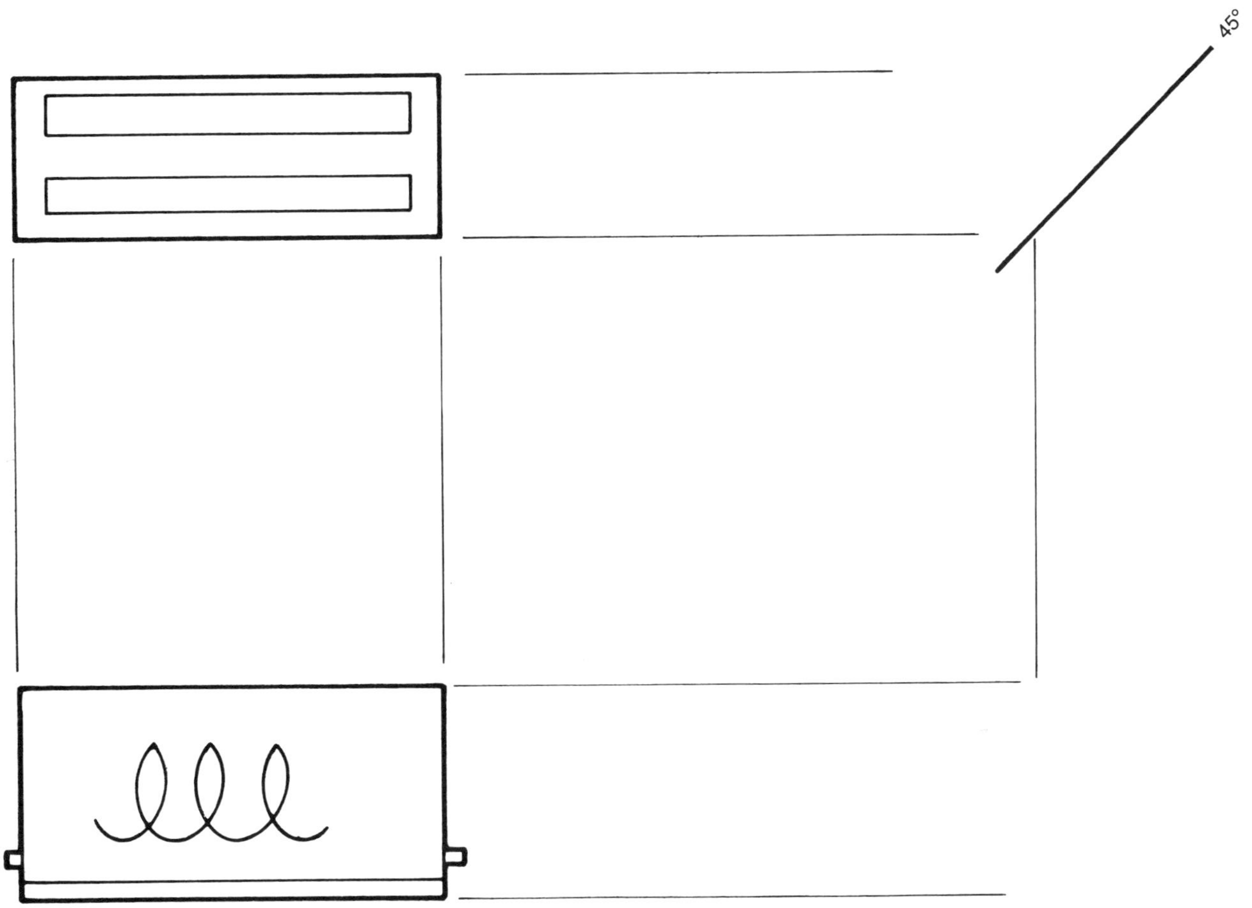

Perspective

Oblique

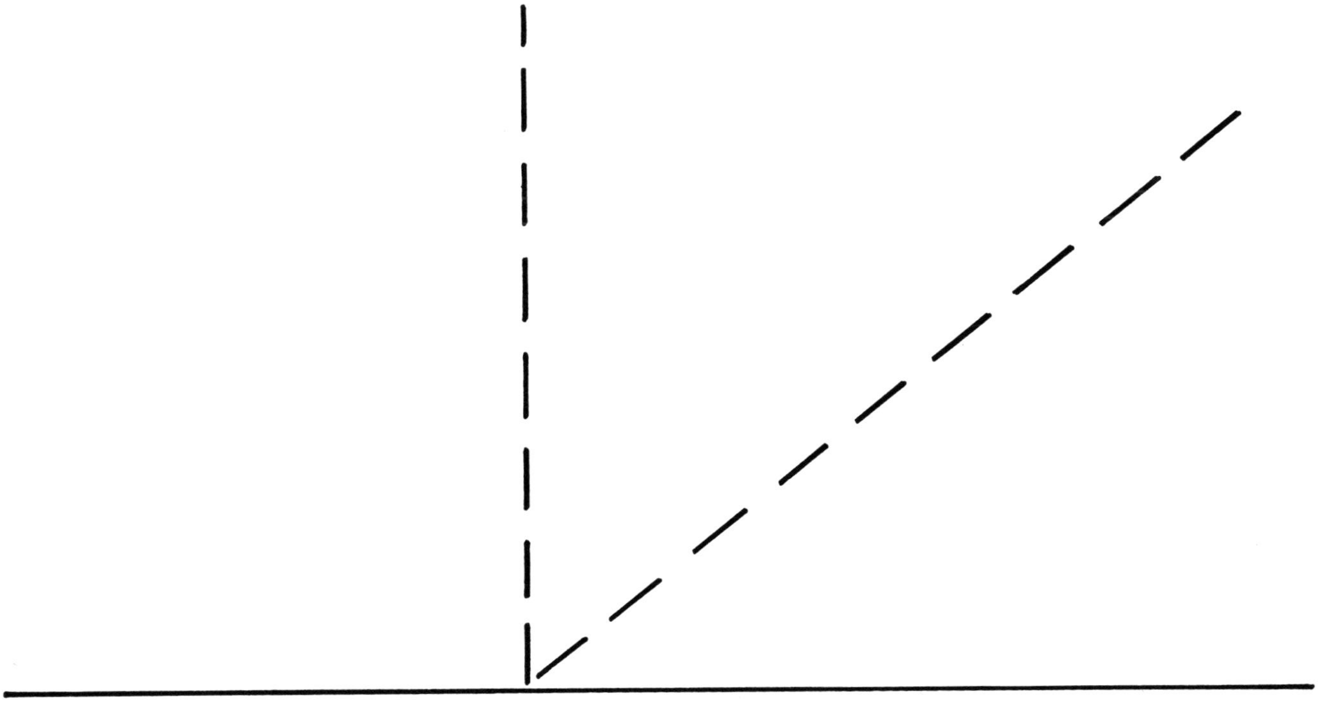

Isometric

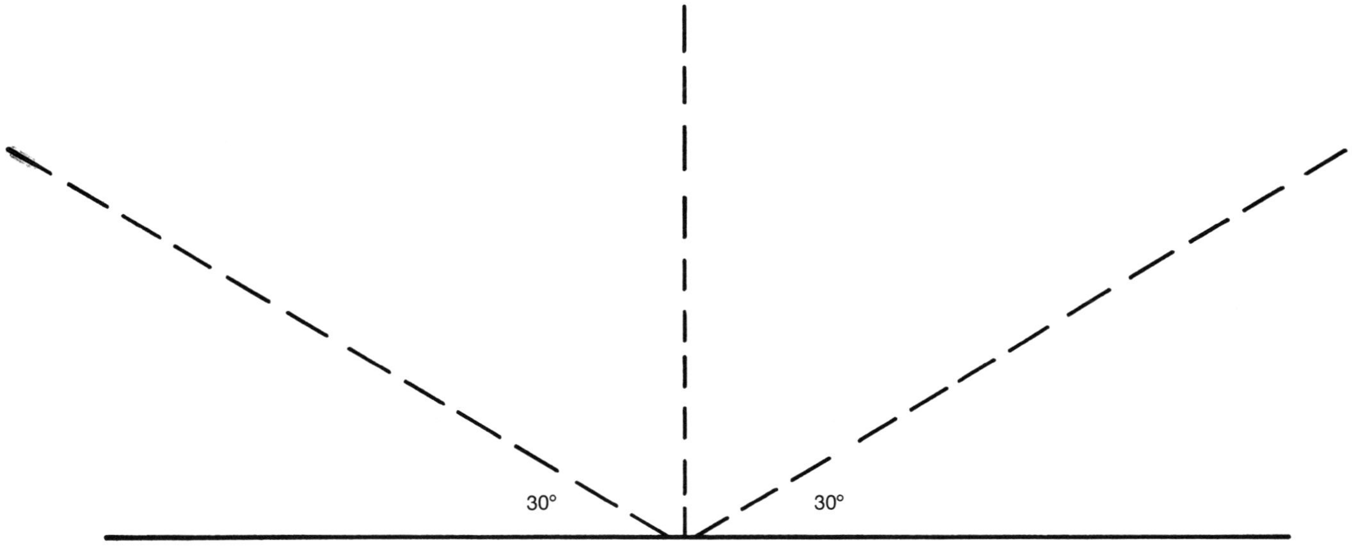

Name	Group Members
Selected Item	

Activity I-8
Hidden Lines

So far all of our drawings have used solid lines to show the edges of objects. Sometimes it is necessary to show edges that are a hidden view. For example, in the top view of the drawing of a yo-yo, the center is hidden from view. It is shown with a dashed line, called a hidden line.

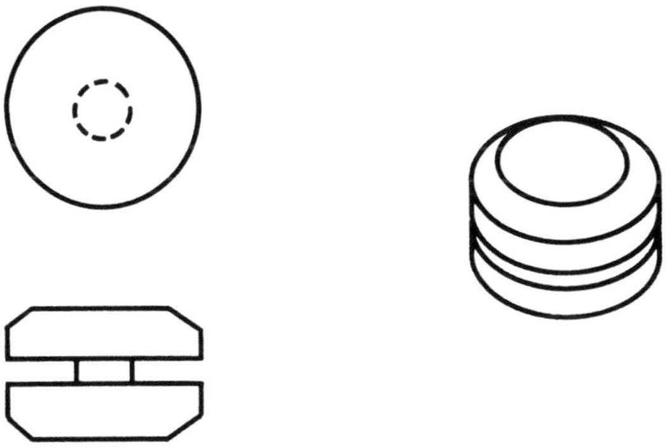

WHAT YOU ARE TO DO:

Sketch the hidden lines for the objects shown below.

A

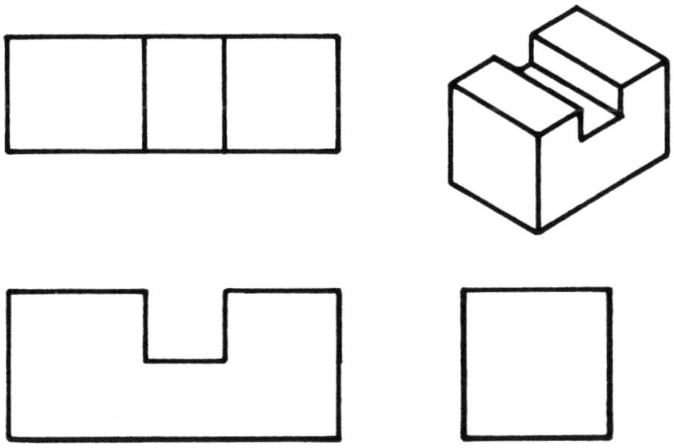

B

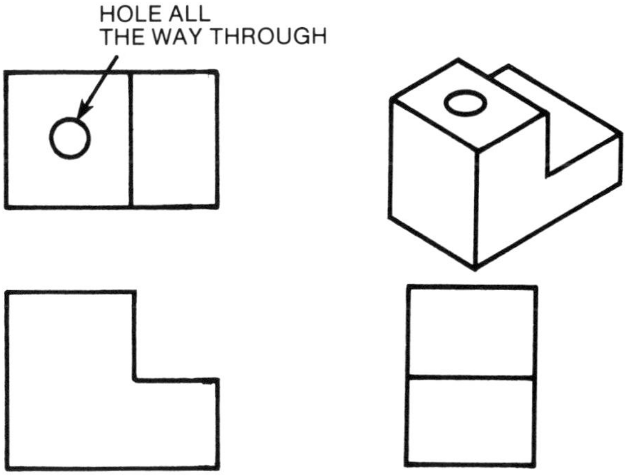

Name	Group Members
Selected Item	

Activity I-9
Dimensions

When an object or device is to be produced, the design must first be done. This design must show the size and shape of each part. To show these, either a three-view or a pictorial drawing can be used. In both cases, the size of the various shapes must be stated. To show the sizes or measurements, dimensions are added to the drawing.

The three-view drawing below consists of dark object lines with medium lines extending from them. These are referred to as *extension lines. Dimension lines* are added between the extension lines and the true measurement is placed. The arrowheads point to the extension lines.

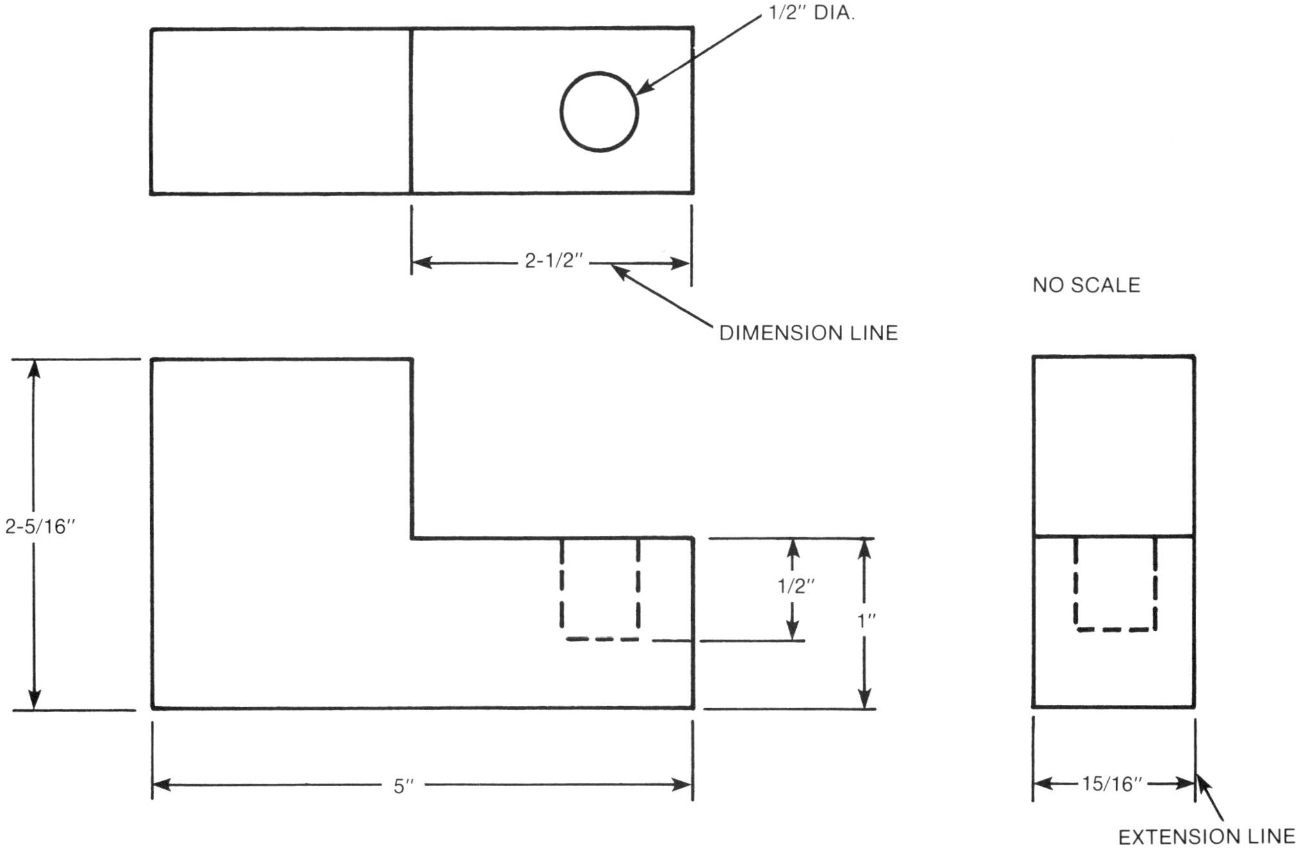

WHAT YOU ARE TO DO:

Turn to the pictorial drawing on the next page and add the dimensions needed to make the item shown.

Dimensions Worksheet

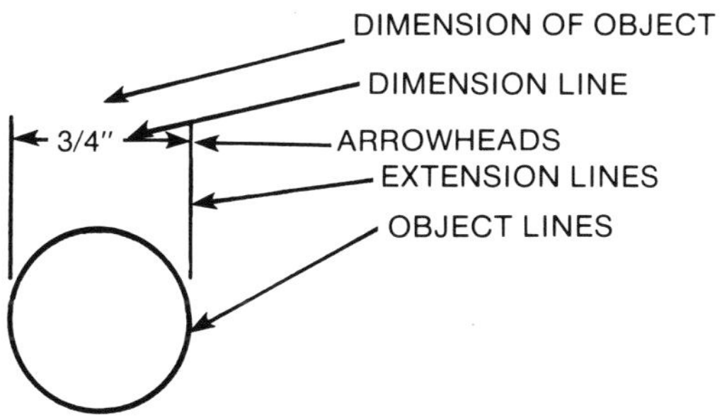

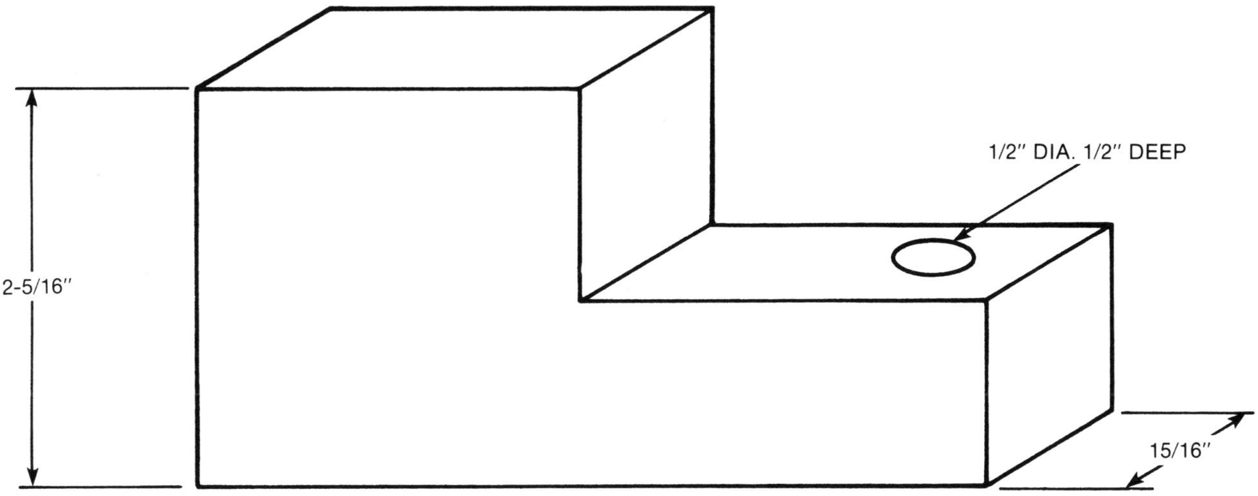

Name	Group Members
Selected Item	

Activity I-10
CAD—Computer-Aided Design

Computers are now used widely in the drafting and design area. There are many types of computers and computer software that you can use for this purpose. Your teacher will demonstrate the computer in your lab so you can see how it is used in the design industry.

Student Using the Computer for a Design

WHAT YOU ARE TO DO:

1. Follow your teacher's demonstration.
2. Try your hand at the CAD system.

Name	Group Members
Selected Item	

Activity I-11
Design of the Product

As mentioned, the design of your project is of extreme importance. When your group has determined which phase is to be represented by the model, you will begin designing.

Examples of Models Built by Students

The drawing on the next page shows what the design should look like, complete with dimensions and a drawing of every part.

Start your design on the following worksheets.

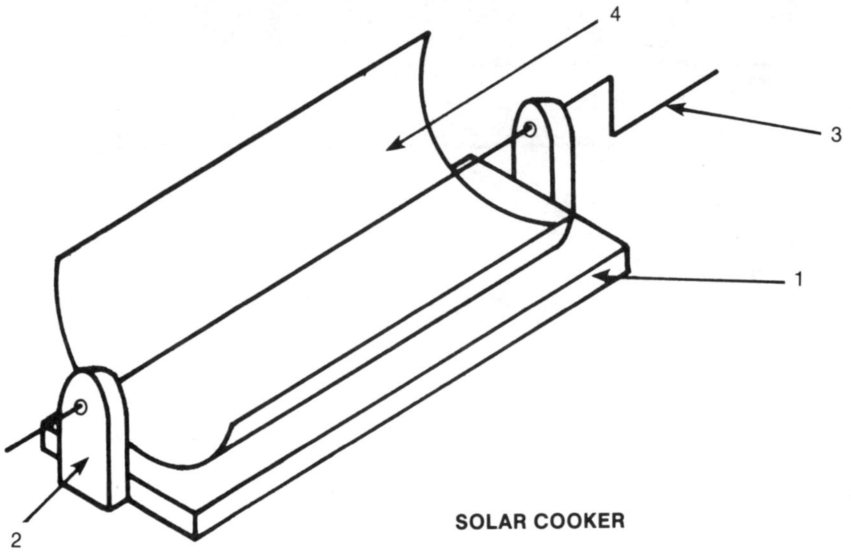

SOLAR COOKER

NO SCALE

PART 1

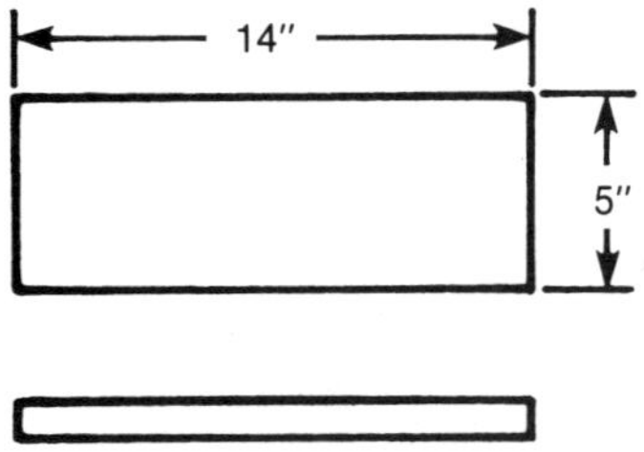

MATERIAL—3/4″ PINE

PART 2

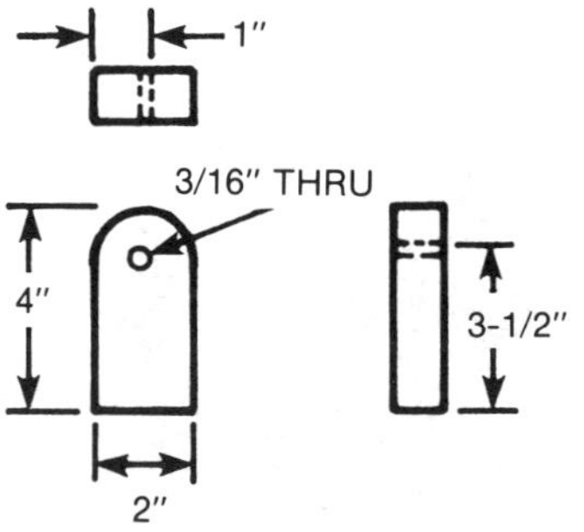

MATERIAL—3/4″ PINE,
2 NEEDED

PART 3

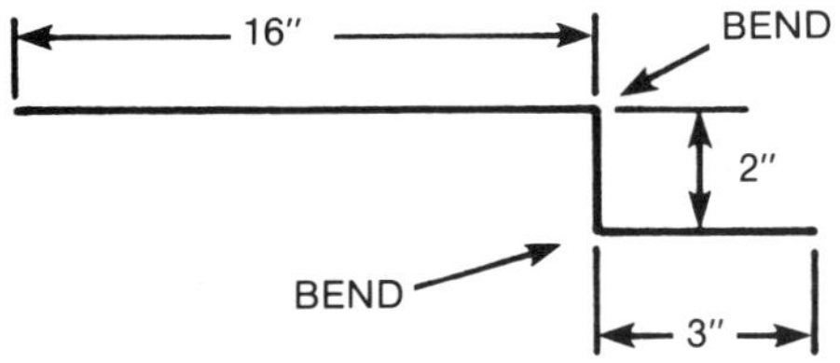

MATERIAL—1/8″ DIA.
WELD ROD

PART 4

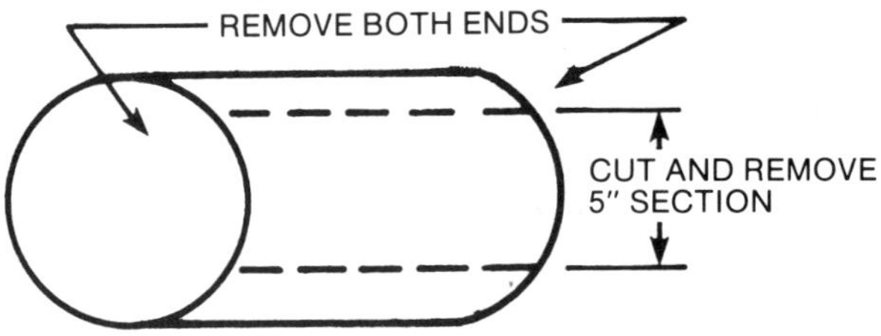

MATERIAL—#3 COFFEE CAN

Design Worksheet

Design Worksheet

Design Worksheet

Name	Group Members
Selected Item	

Activity I-12
Use of Tools

As you go from the design phase of the product to actually building it, you will need to use the basic hand tools and electric tools available in your school.

Your instructor will guide you in the proper use of these tools. Filling out the chart found on the next few pages of the book will give you an easy source of reference.

Note: Safety is of extreme importance and must be considered even when using the simplest hand tools. Always review the safety entry in your tool chart before you begin. Be sure safety glasses are worn when necessary and that you are wearing the proper clothes for safe working. If you have any problems, ask your instructor for help!

Tools Worksheet

As your instructor covers the tools in the lab, write their names and purposes below.

TOOL NAME | **PURPOSE/USE/SAFETY**

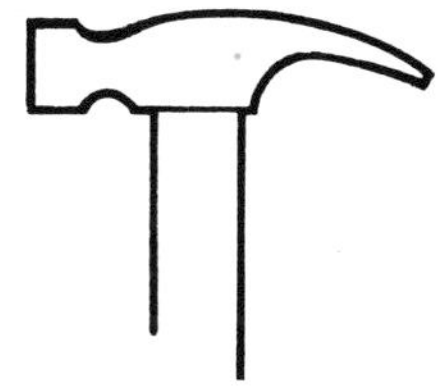

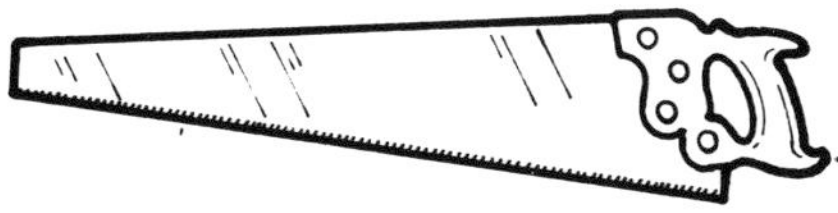

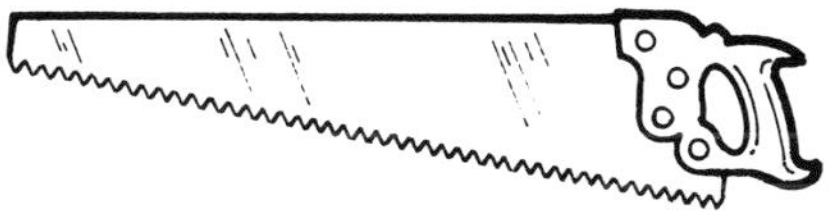

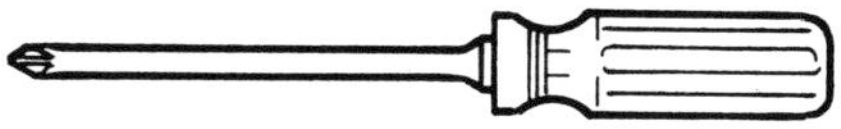

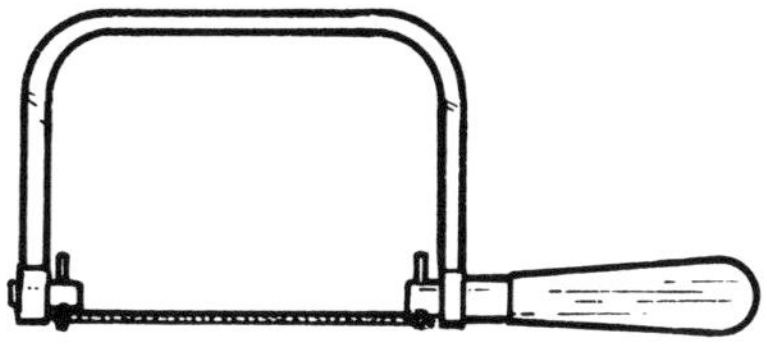

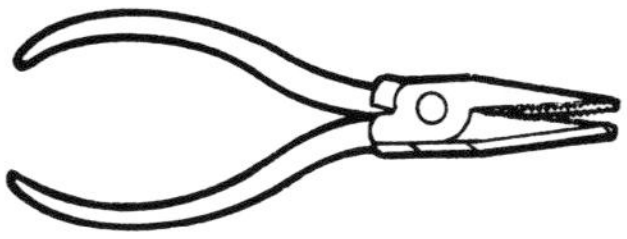

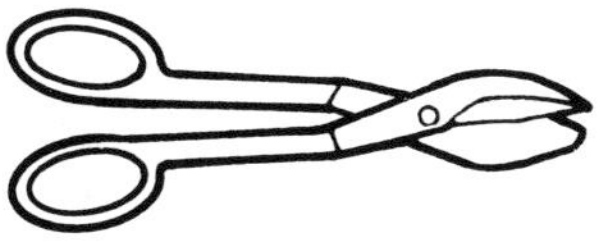

	TOOL NAME	PURPOSE/USE/SAFETY

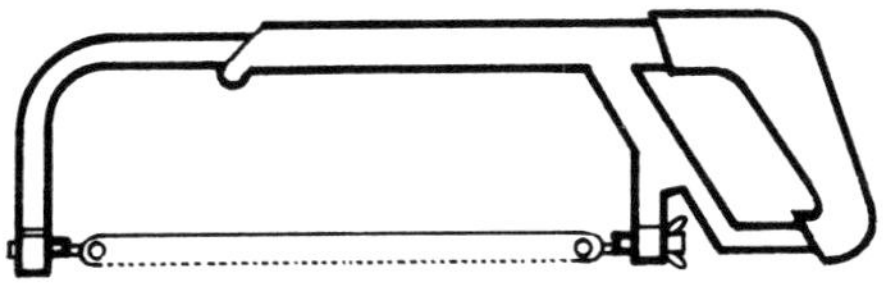		
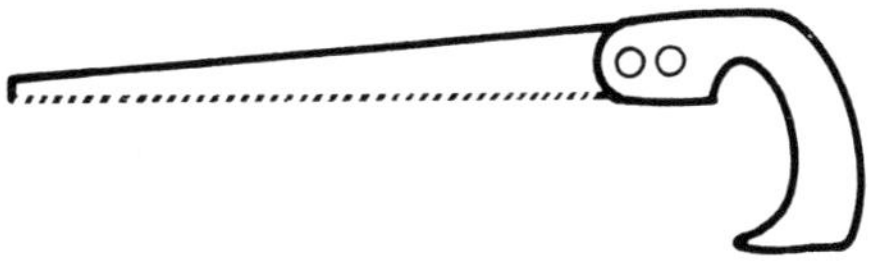		
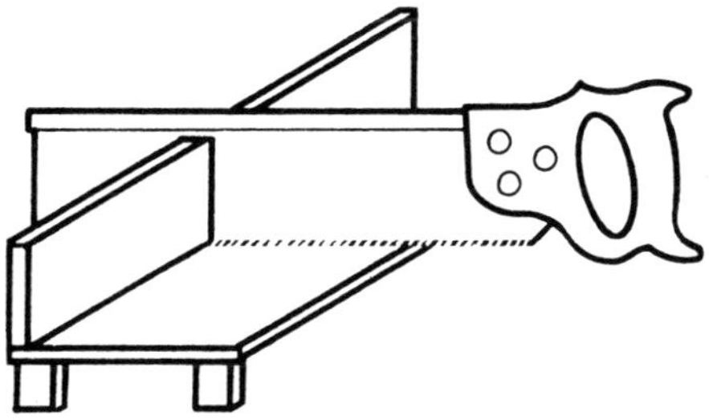		

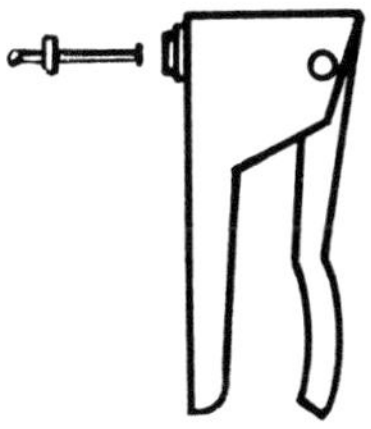		
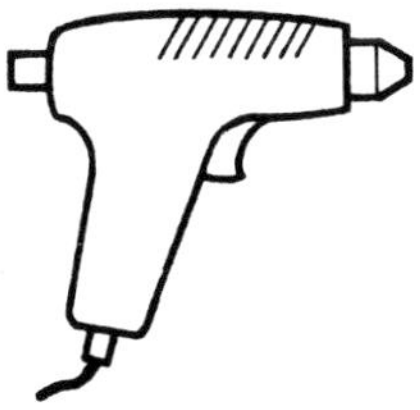		

TOOL NAME	PURPOSE/USE/SAFETY

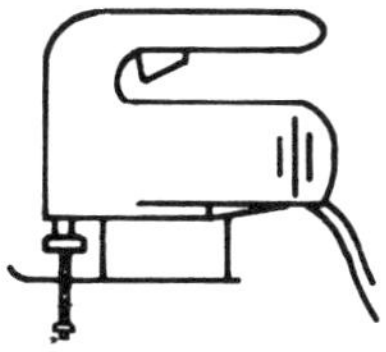

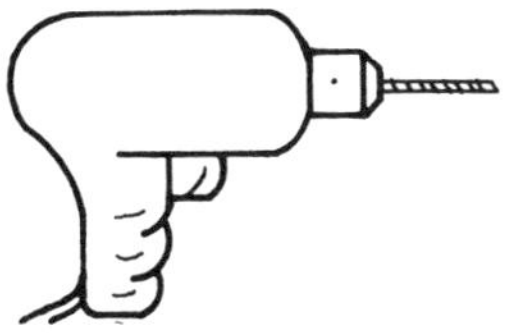

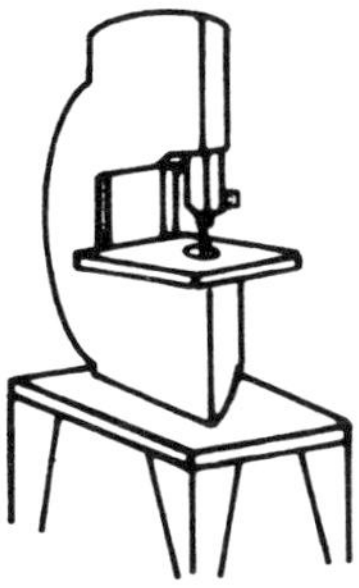

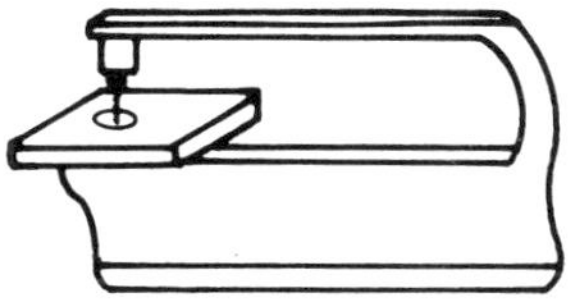

Name	Group Members
Selected Item	

Activity I-13 Using the Camera

The camera is a device that most people have used at least once in their lifetime. Still, many people make mistakes while using the camera.

Many of the activities in technology education involve the use of photography. In fact, the camera is a valuable tool for design and for recording the progress and results of your projects. Learning to use a camera and to develop and print your own pictures can be very rewarding—and fun, too.

There are many different types of cameras in the world today, but all have the five following parts:

1. All cameras have a *"light-tight" box.*
2. The *lens* and *shutter* allow measured amounts of light to reach the film. The lens focuses the image onto the film, while the shutter controls the light entering the camera. (See page 67, Part 3)
3. The *shutter button* is the device pressed to open the shutter (Part 4).
4. The *viewfinder* is what you look through to see the subject being photographed. The viewfinder also lines up the subject with the camera's lens (Part 2).
5. The *film advance* moves the film inside the camera so that a fresh piece of film is always behind the lens (Part 1).

WHAT YOU ARE TO DO:

1. Fill in diagram on the next page.
2. Use the camera for one of your phases on either the Evolution Board or the How Things Work Board.

When using the camera remember these four important facts.

1. Make sure you have the proper amount of light.
2. When pushing the shutter button make sure the camera does not move.
3. Check to make sure the focus is correct.
4. Look behind the subject and make sure there is a good background.

Camera Parts

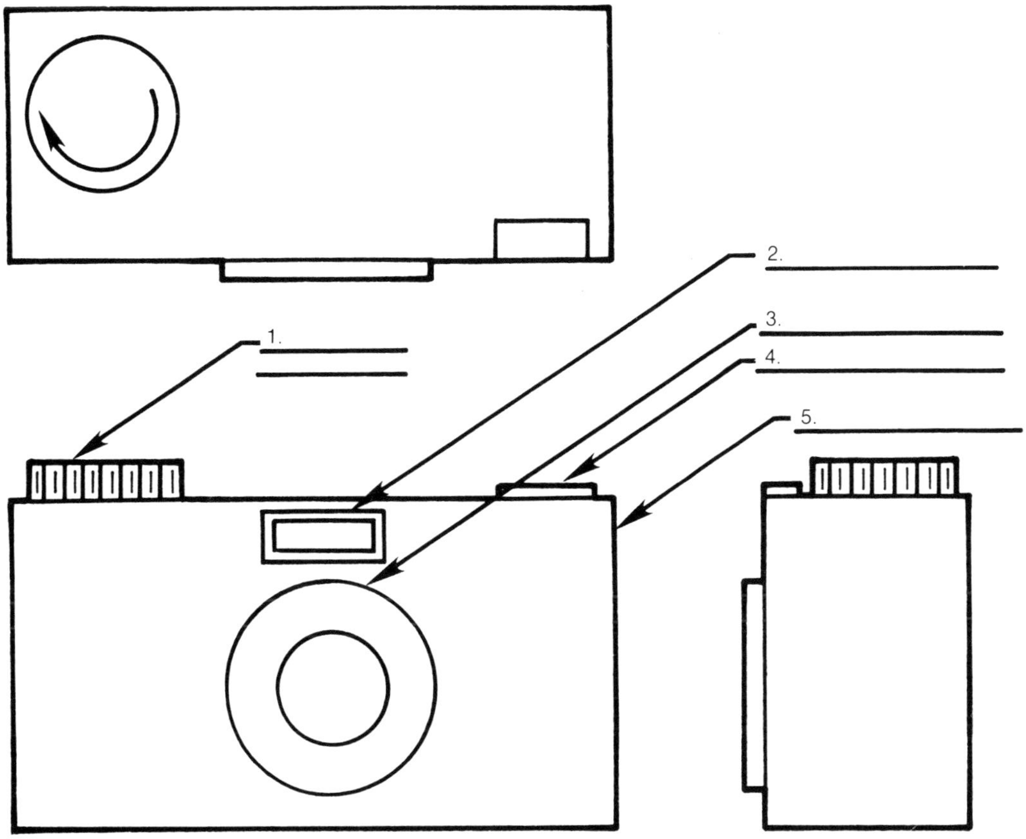

WHAT HAPPENS INSIDE THE CAMERA?

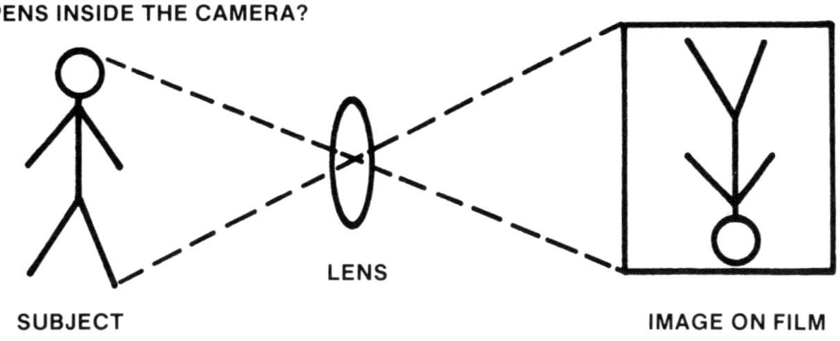

Name	Group Members
Selected Item	

Activity I-14
Developing Black and White Film

Developing this film is a simple procedure. Once your film has been exposed, the images created on your film by the light reflected from the subject must go through a series of steps in order to become a film negative. In a few of these steps, chemicals are used. The chemicals and their purpose are:

1. *Film Developer* brings out the image.
2. *Stop Bath* stops the developer from working.
3. *Film Fix* makes film no longer light sensitive and preserves the negatives.

WHAT YOU ARE TO DO:

1. Read the direction sheet. Get all the materials, *developing tank, apron,* and *weight.* Arrange materials so you can find them in complete darkness.
2. In total darkness, remove the film out of the cartridge. Roll the film into the *film apron.* Set the film and apron in the developing tank, set weight on film, and cover the developing tank. The developing tank is now in the darkroom. Remaining steps may be carried out in the light.
3. Pour developer into the tank. Leave in the developer for approximately 8 minutes (check any photographic chart for exact time).
4. Pour stop bath into the tank; leave in for 20 seconds.
5. Pour film fix into the tank. Leave fix in tank for 8 minutes.
6. Remove tank cover and fill tank with water.
7. Allow *negative* to remain in rinse for approximately 20 minutes.
8. Hang negative and allow it to dry.

Name	Group Members
Selected Item	

Activity I-15
Making Black and White Prints

The *enlarger* is a device that will be used to make all sizes of photographs. It works much like a slide projector. The farther away a slide projector is from the screen, the larger the image that is projected. The closer the projector is moved toward the screen, the smaller the image will appear. In the enlarger, the farther the enlarger is away from the easel, the larger the print that will be made. Once your negatives are ready, set the negative you wish to enlarge in the negative carrier and follow the steps listed below.

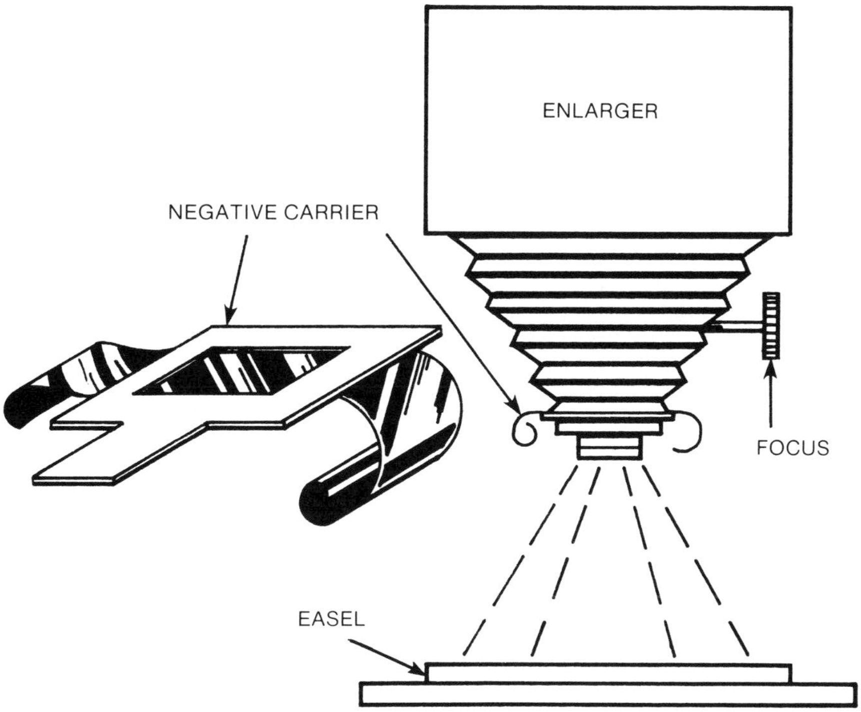

WHAT YOU ARE TO DO:

1. Adjust the height of enlarger to give the size print desired.
2. Focus the enlarger so a sharp, clear image is projected.
3. Arrange the easel so that it is in the desired position.
4. Set the photographic paper in the easel when the lights are turned off (shiny side up, emulsion side).
5. Set the timer to the desired time and turn on the enlarger.
6. Remove photographic paper from the easel and develop it.

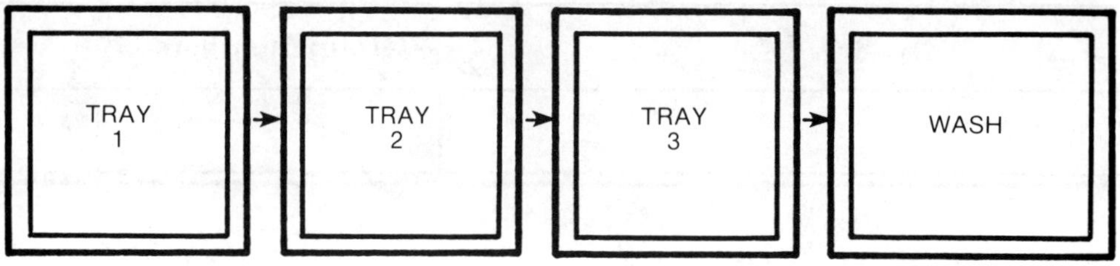

7. Tray 1 contains *paper developer.* Place the exposed paper in the developer until the desired image appears.
8. Tray 2 contains *stop bath.* Place the photograph in it for about 10 seconds.
9. Tray 3 contains *fix.* Leave the photograph in it for 3 minutes.
10. In the last tray, wash the photo, wipe it, and allow it to dry.

Name	Group Members
Selected Item	

Activity I-16
Using the Technology Time Line

Completing your Evolution Board and How Things Work Board projects will show how technology has always existed in history and how it has changed. Using a technology time line is a good way to visualize the growth of technology and to see where your project fits in. The time line also can help show how some technologies make later ones possible; for instance, the discovery of electricity in the 1700s made possible the invention of batteries, light bulbs, record players, and even computers.

The computer is a good tool for recording time line information since new entries can be made without having to rewrite the time line.

WHAT YOU ARE TO DO:

1. As your group discovers the dates of their items, they should be placed on the time line.
2. Input the information to your class computer.
3. As other groups report their findings, have all the dates set on the time line found on the next page.
4. Be able to identify and discuss the importance of the time line entries.

Technology Time Line

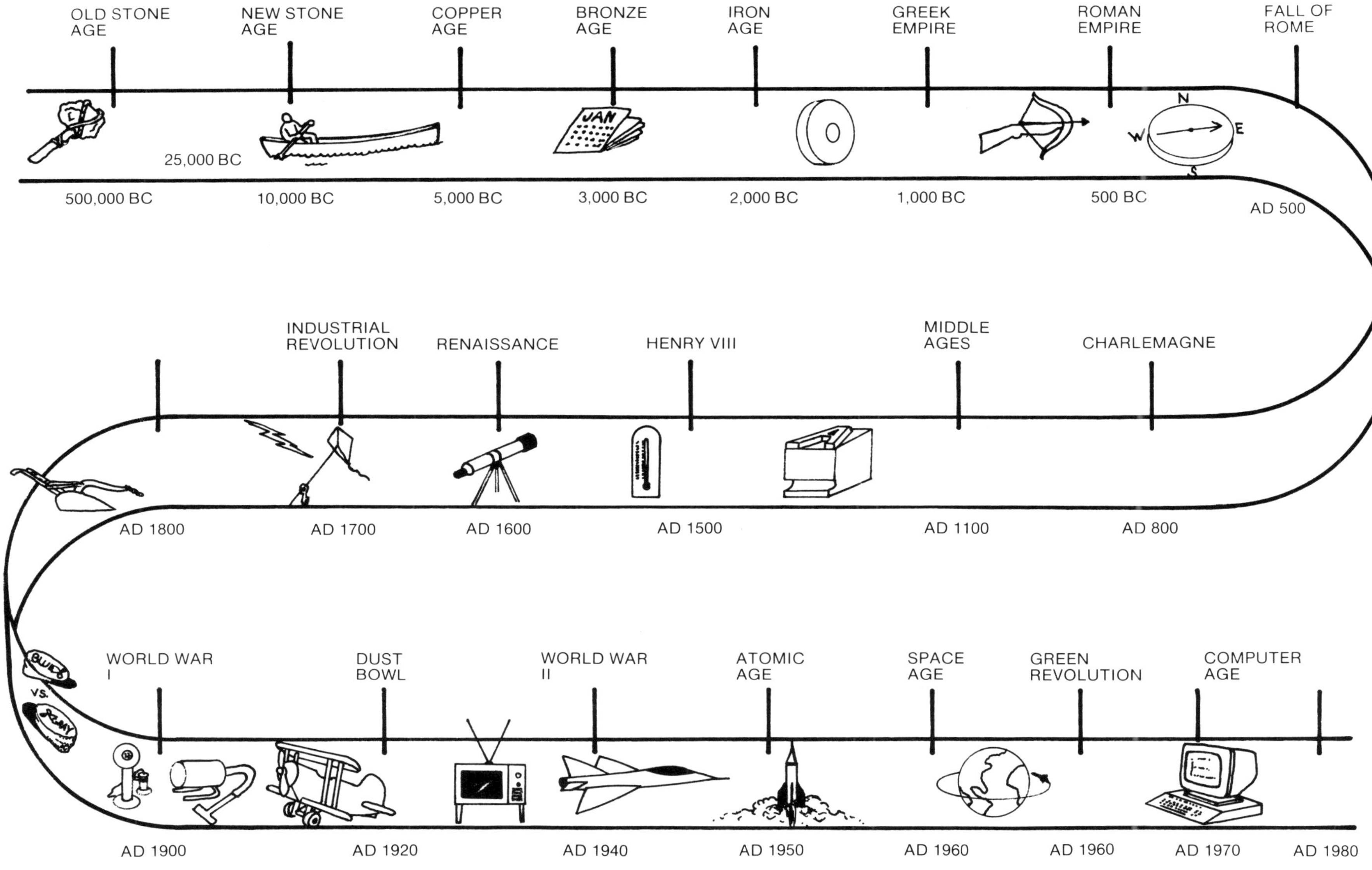

SAMPLE TIME LINE ENTRIES

Device Invented	Year/Time Line	Inventor (If Known)
Abacus	3000 BC	
First sailing ship	1200 BC	
Amber and its ability to attract materials	600 BC	Thales
Sundial	AD 984	
Hand fan	AD 700	Japan
Camera obscura	1039	
Guitar	1500	
Food cooler	1600	
Electricity named	1600	Gilbert
White light	1666	
First battery	1800	Volta
Magnetic field in wire	1819	Oerstead
Electric motor	1820	Faraday
Electric waves in space	1825	Hertz
Telegraph	1836	Morse
Telephone	1876	Bell
Electric light	1879	Edison
Power plant lit electric lights	1882	
Fountain pen	1880	
Steam washing machine	1884	Morton
First transformer	1895	Marconi
Wire vacuum tube computer	1944	
One step photography	1947	
Transistor computer	1956	US Air Force

Unit I Evaluation

Name ______________________ School ______________________

What is technology? ______________________

____

____

List the three aspects of technology:

____

____

____

What was your activity? ______________________

What was your device? ______________________

List your phases or how item worked. ______________________

____

____

What did you place on the time line and dates? ______________________

____

____

How did (or does) this device do the following?

Meet need or want ______________________

____

Influence life ______________________

____

What aspect of technology (type of technology) does it show? ______________________

Why? ______________________

GROUP MEMBER	RATING OF MEMBERS		
	Good	Poor	Fair
______________	☐	☐	☐
______________	☐	☐	☐
______________	☐	☐	☐
______________	☐	☐	☐

UNIT II

Resources

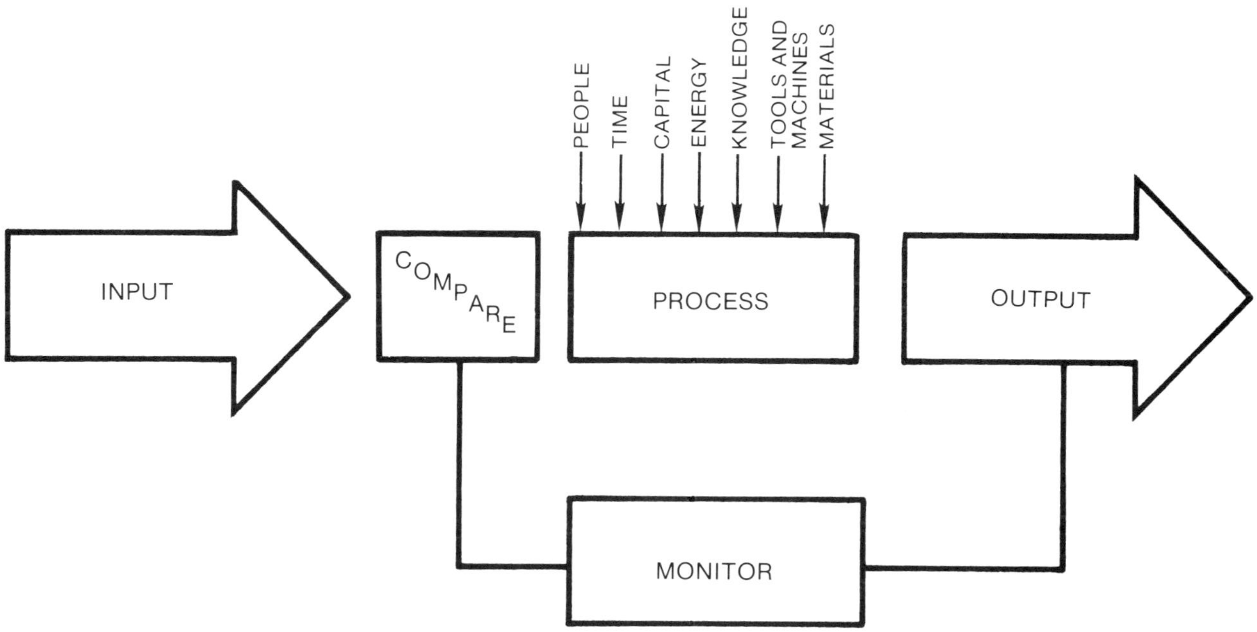

Technology needs and uses *resources* in order to carry out desired processes. This is true whether or not the technology involves the making of a hamburger or sending a spacecraft to other planets. In each case the seven basic resources of technology can be found.

The following six activities will introduce the various resources and show how they fit into the world of technology.

Name	Group Members
Selected Item	

Activity II-1
Seven Resources of Technology

WHAT YOU ARE TO DO:

1. Define each resource in your own words.
2. Give an example of each as it relates to technology.

a. People—____

____

____

b. Time—____

____

____

c. Tools and Machines—

d. Energy—

e. Materials—

f. Knowledge (how it is different than information)—

g. Capital—

Name	Group Members
Selected Item	

Activity II-2
Technology Process Chart

Tools and machines make up one of the basic resources of technology. They are each designed to carry out a particular part of the technology process, and it is important to understand how this is done.

WHAT YOU ARE TO DO:

As you use a tool or machine in your activities and projects, or carry out a process, fill out the Technology Process Chart. It will be a good reference for finding what tools and machines are involved in the various processes of technology.

TECHNOLOGY PROCESS CHART

Layout and Measurement	Cutting and Machining	Forming and Molding	Fastening	Coating and Finishing

Name	Group Members
Selected Item	

Activity II-3
Processing Resources

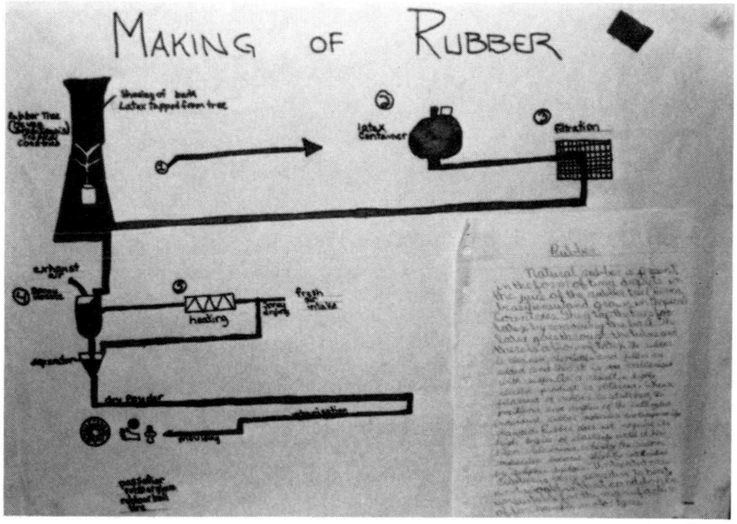

WHAT YOU ARE TO DO:

Your instructor will pass out a material resource card. Your group will receive one of these cards that contains the name of one material. Your group will make a detailed drawing showing how that material (or one of its specific uses) is processed.

Along with a drawing, there must be a list that states some of the material's most common uses, the type of material it is, if it is renewable or nonrenewable, location of the raw material, and if the products obtained from it are recyclable or not.

The following is a basic list of materials that your selection can come from: plastic, textile, rubber, ceramics, wood, metal, paper.

Report findings to the class.

Be sure you fill out the study sheet on page 93 when other groups report their findings.

Your Group's Material ______________________________

Group Members ______________________________

____

____

Use this page for your group's sketch of the materials process. Your group is to do a neat, detailed drawing.

MATERIAL PROCESS STUDY GUIDE

Material	How a material is processed, the raw material it comes from, and the material's uses.
Plastics	
Textile	
Rubber	
Ceramics	
Wood	
Metal	
Paper	
Other	

Name	Group Members
Selected Item	

Activity II-4
Energy Resources

Model of Oil Well Showing Where Oil Is Found and How We Use It

There are many different forms of energy, some of which have been used by humans for ages. Through technology, humans have developed methods through which newly discovered energy resources can be used. In either case, energy resources help humans meet their needs and wants. Energy resources can be either renewable or nonrenewable in nature. Some specific energy resources are: wind, water, coal, natural gas, nuclear, solar, and geothermal.

In this activity your group will be given one of the resources. You will design and build a device (a model or an actual working device) that once used, currently uses, or will use your energy resource in the future. An information sheet will state where the source is located, how the source is used to meet human needs and wants, and whether or not the source is renewable or nonrenewable. Each group will report all findings to the class. Information will be recorded by your computer. Make sure your report answers all the questions on the energy resource study sheet.

Your Group's Energy Resource ______________________________

Group Members ______________________________

Be sure to fill out a study sheet when other groups report.

Use this page for a drawing of your group's energy source device. Groups using the lab's drafting and design materials are to design their device.

ENERGY RESOURCE STUDY GUIDE

Resource	Device	When device was used (past, present, or future), what device does, and whether or not the resource is renewable.
Wind		
Water		
Solar		
Coal		
Oil		
Natural Gas		
Nuclear		
Geothermal		

Name	Group Members
Selected Item	

Activity II-5
Material Testing

The world consists of many materials. These materials can be used in the making of products. The products are designed and built for a specific human need or want. The proper selection of materials is extremely important to the designer. All materials have various properties that deal with strength, durability, weight, brittleness, ease of use, and flexibility, to name a few. These properties play a vital role when selecting the materials to be used.

WHAT YOU ARE TO DO:

1. Your group will be given a card with a testing method on it. You will discover what the method is, then design and build a simple testing device. Your group will then perform tests on a material you choose.
2. Report to the class the definition of the testing method you chose, and report the results of the test you performed.

Group Testing Method ______________________________

Group Members ______________________________

Testing methods for various properties are shown on the following pages.

Material Testing Ideas

TENSILE TESTING

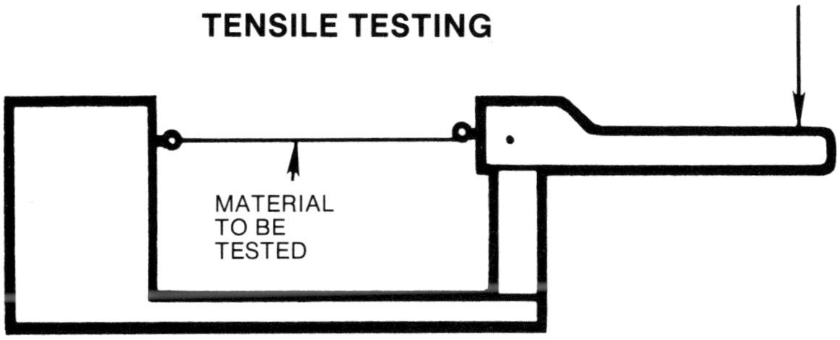

COMPRESSION TESTING

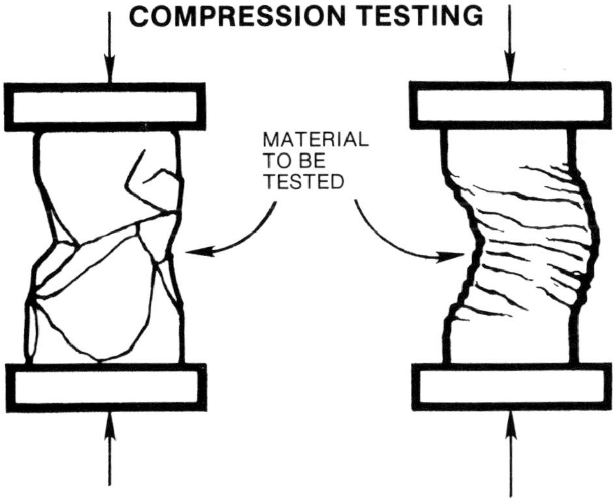

CONDUCTIVITY TESTING

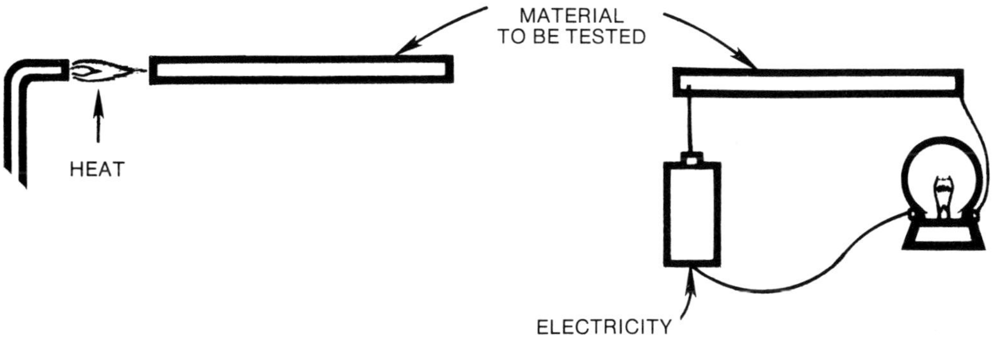

ABRASION TESTING

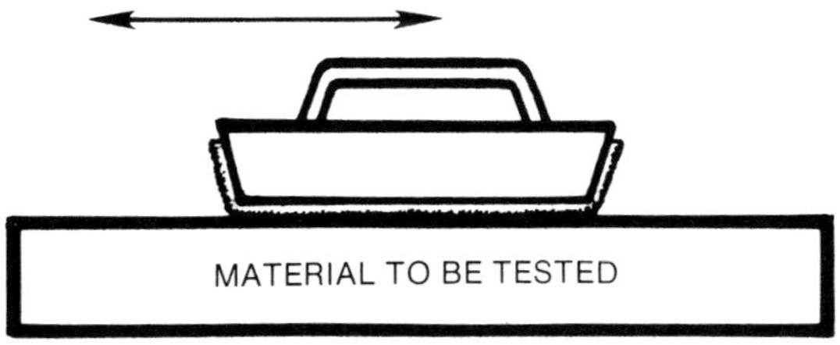

HARDNESS TESTING

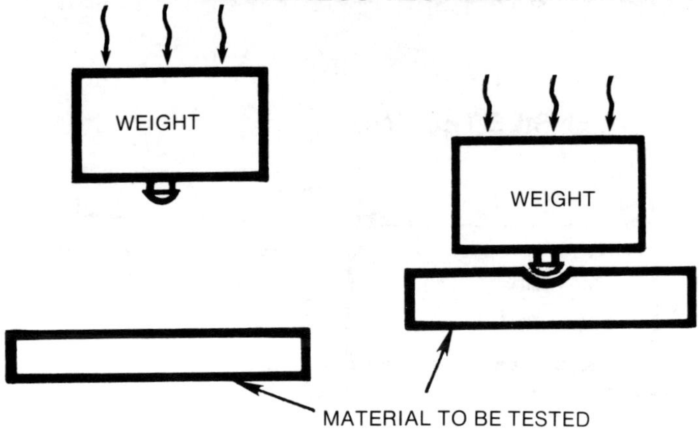

ELASTICITY TESTING

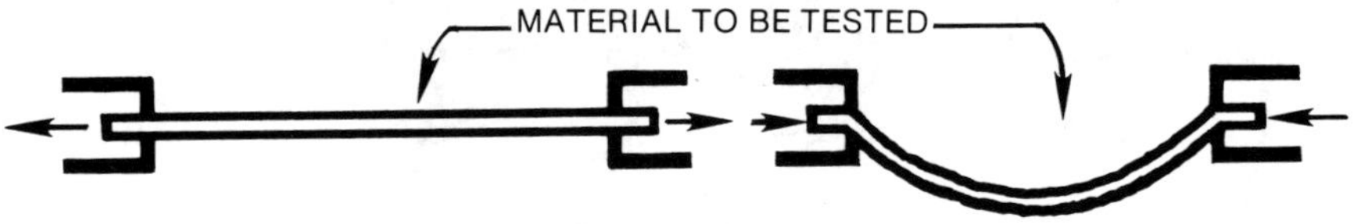

ADHESION TESTING

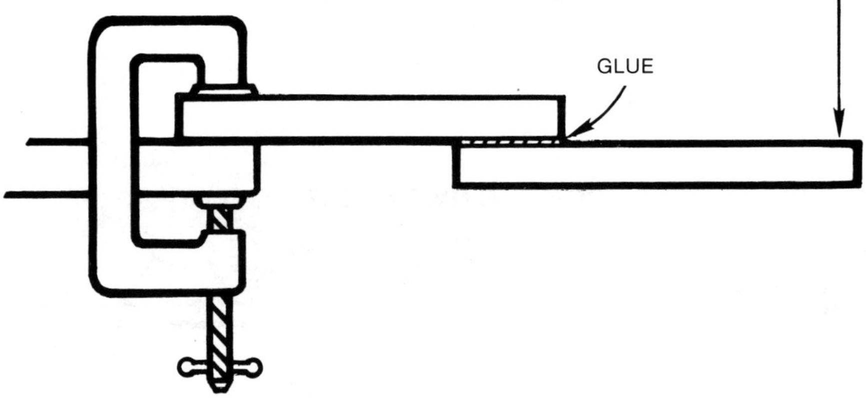

Make a drawing of your group's testing method.

TESTING METHODS STUDY GUIDE

Testing Method	Definition of term and possible materials term could relate to:
Tensile	
Compression	
Hardness	
Abrasion	
Adhesion	
Elasticity	
Conductivity	
Other	

Name	Group Members
Selected Item	

Activity II-6
Location of Resources

Advances in communications and transportation have helped make our world "smaller." Raw materials from all over the globe are now used to produce the products we need and want. Oil is drilled in vast deserts, the Arctic Circle, and the ocean depths by offshore oil rigs. Other materials are found in jungles, forests, and mountain ranges. Each part of the world has certain resources for modern technology.

WHAT YOU ARE TO DO:

1. Select three areas of the world (Texas, Taiwan, and Brazil, for example). Make sure they haven't been selected by other class members. Find out what resources they are known for.
2. Write the name of the resource on a map flag, as shown below. The flag will be placed on the world resource map to show what state or nation it comes from.

Resource		Area of the World
______________________		______________________
______________________		______________________
______________________		______________________

3. Report your findings to the class.
4. Add your area and its resource to the computer.
5. Complete your Resource Location Study Guide.
6. On your map, locate five areas and write down their resources.

RESOURCE LOCATION STUDY GUIDE

Resource	Location
Oil	
Coal	
Natural Gas	
Nuclear	
Geothermal	
Rubber	
Ceramics	
Plastics	
Wood	
Metal	
Paper	
Textile	

If groups report certain types of the above resources, be sure to list them under the correct resource title.

Location of Resources on World Map

Place a ★ next to the location with the resource(s).

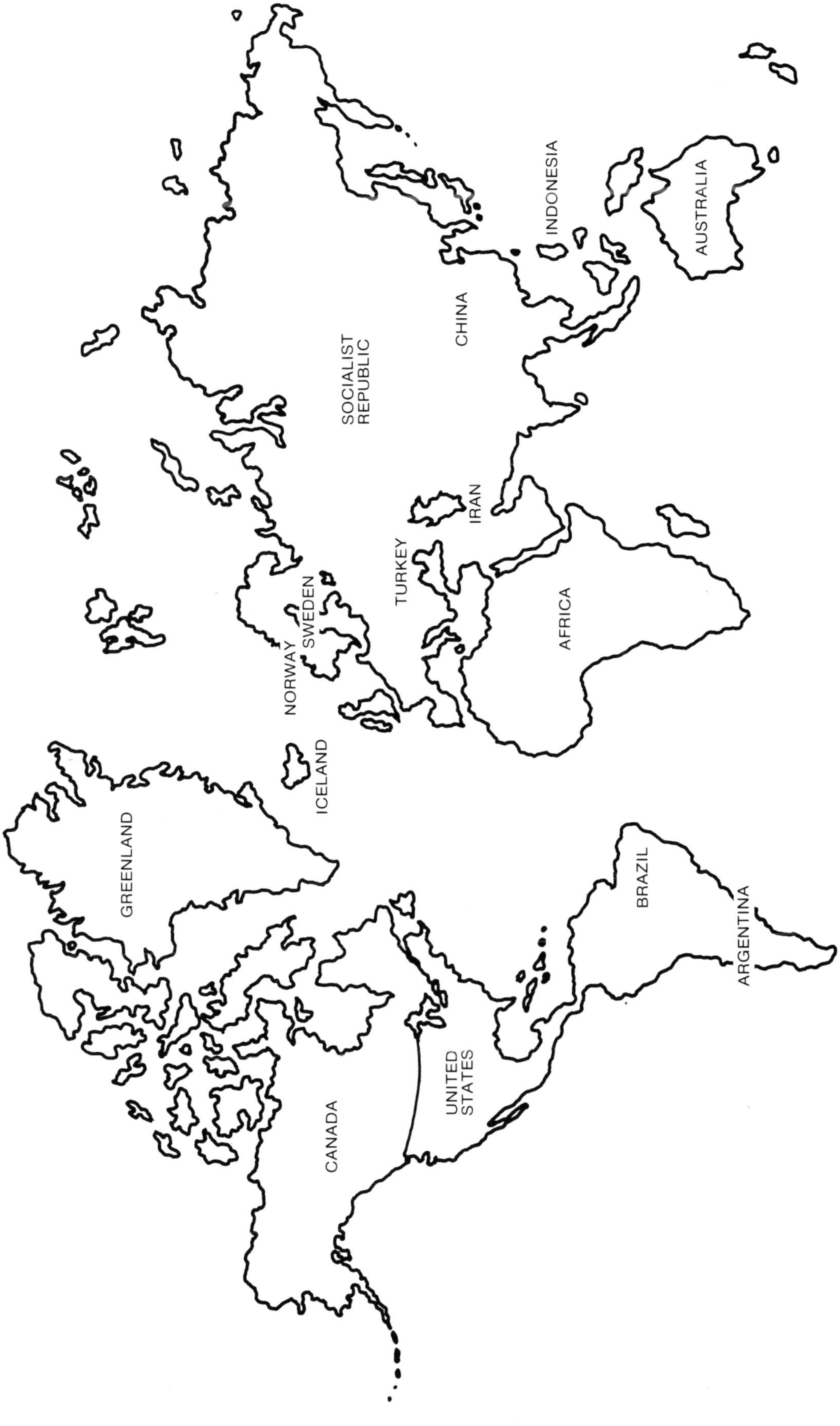

UNIT III

Problem Solving

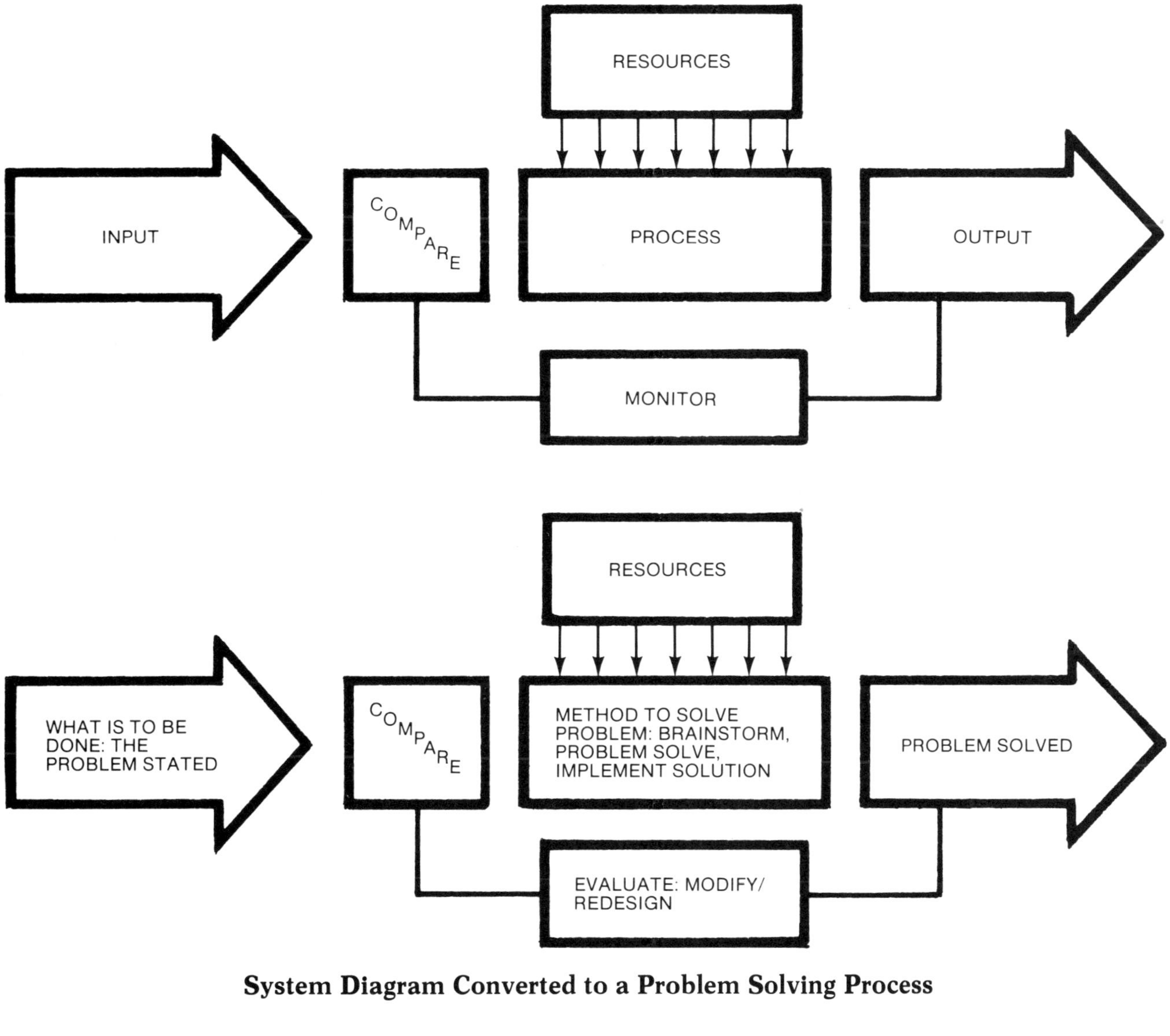

System Diagram Converted to a Problem Solving Process

By now, you should know that technology can be defined as *using resources to meet human needs and wants.* In Unit I, we learned and discovered what technology is and how it has changed our world.

Unit II concentrated on the resources of technology. As you should know, there are seven resources. They are: *time, people, knowledge, capital, tools and machines, energy,* and *materials.*

Whenever there is a need or want, there is a problem that must be solved. One of the methods of problem solving is to list all possible solutions and to select the best solution from that list. This first method is called *brainstorming.*

Students in a Brainstorming Session

After the brainstorming is performed, the problem is ready to be solved. There is a formalized method, or system, used to solve a technological problem. The system we are going to use consists of seven steps. These seven steps are:

1. Define the problem
2. Identify the parameters
3. Survey the available data
4. Develop an hypothesis
5. Select a solution
6. Implement the solution
7. Evaluate, redesign, and modify if necessary

Let's take a closer look at these seven steps of problem solving. First, look at the problem and determine what it really involves. Find out what the facts of the problem are and when necessary, list these facts. This step is referred to as *defining the problem.*

To *identify the parameters,* it is necessary to know and list all the limits of the problem. Once the problem is defined and the parameters are known, look carefully at your list and then make a few educated guesses (assumptions) on how you could solve the problem. This educational guess is what is known as the *hypothesis.* The next step, *selecting the solution,* is deciding what's to be done to solve the problem.

Now that the solution is selected, go ahead and *implement the solution.* In other words, begin the task of solving the problem.

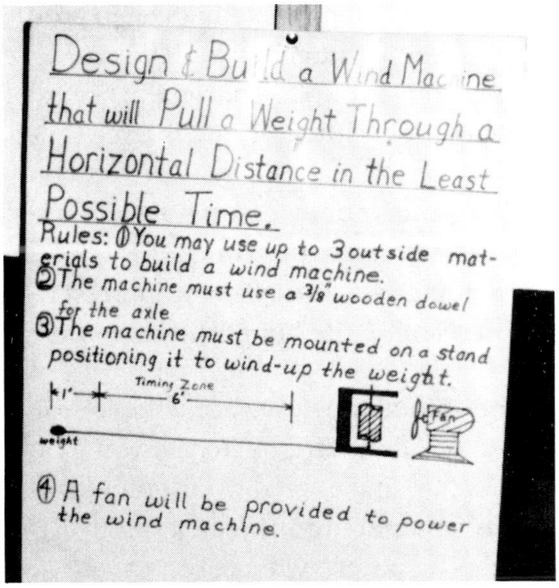

The Problem

The Solution

The problem is now solved—or is it? Look at your solution. Are there modifications that can be carried out to make your solution better? If so, *redesign and modify* to meet the desired solution.

Students Presenting Their Solution to Coin Shortage Problem

Note: You will be required to solve one of the problems stated in this unit, or one stated by your instructor.

Before you begin your problem solving activity, there are a few facts you must know.

A *technology log* must be kept by each group member. In this log you are to record what your group did during each class period. This log is where you will keep track of the *technology money* spent and all charges that were made to rent the various resources allowed.

Your group will start with a certain amount of money, usually $36.00 per person in each group. This money will be used for the rental of tools and machines, and some materials. See the resource rental list in the appendix. The *charge card* provided, will enable your group to charge the rental of the resources needed. An *interest rate* will be applied to the charge. This interest rate is a fee that is charged to you for using the credit allowed.

A *savings deposit* and *withdrawal slip* are also provided so your money can be placed in the technology bank. The money deposited in the bank will draw *interest* for your group. This time the bank will pay you money for the money in your account. The instructor will establish the interest rate in both the cases mentioned above. Your instructor will also determine the hours the bank will be open.

After the problem has been stated and your group has determined how they are going to solve the problem, they will apply for a *patent.* The patent application will contain a sketch showing how the problem is to be solved. The patent will protect your idea from being copied by another group. Your group will receive a number; this is your patent number. The patent number must be on all the designs made by your group and also on the finished product. The patent will cost the group the amount of $5.00 per group member.

During the evaluation and redesign of the device, keep track of all modifications and their results on the graph provided. The graph will show how the modifications affected your device. Note how the various modifications affected the results of the moving device. Also note how the group stated the modifications in their log.

Log Keeping—
Sample Log Page

NOV. 05, MONDAY
- GROUP COMPLETED LAST CUTS OF DEVICE
- GROUP ASSEMBLED ITEMS – DEVICE COMPLETED
- WOOD GLUE USED TO ADHERE PARTS

SPENT $5.00 FOR CUTS
$5.00 FOR GLUE
TOTAL SPENT $10.00 TODAY
BALANCE OF $46.00

NOV. 06, TUESDAY
- RAN FIRST TRIAL RUN
- DEVICE MOVED FOR 16 SECONDS
- SEEMED TO BE OUT OF BALANCE — GROUP REMOVED POINTED PART AND CENTERED IT ON LARGE WHEEL.
- HOT GLUE USED TO ADHERE POINTED PART

$1.00 SPENT FOR GLUE STICK
$5.00 SPENT FOR GLUE GUN
TOTAL $6.00 SPENT
BALANCE OF $40.00

HAVE 3 MIN. OF GLUE GUN CREDIT.

NOV. 07, WEDNESDAY

RAN RUN #2 – DEVICE MOVED FOR 43 SECONDS
DECIDED TO ADD WASHERS TO LARGE WHEEL – THIS, WE HOPE WOULD GIVE MORE WEIGHT AND CENTRIFUGICAL FORCE – MEANING TO CAUSE ITEM TO SPIN LONGER.

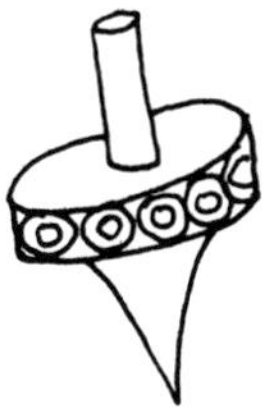

RAN TEST RUN #3
DEVICE MOVED "SPUN" FOR 63 SECONDS.

$3.00 SPENT FOR WASHERS
$1.00 FOR QUE STICKS
USED REST OF TIME ON GLUE GUN
$4.00 TOTAL SPENT
BALANCE OF $36.00

NOV. 08, THURSDAY

- SINCE ADDING MORE WEIGHTS IT DID IMPROVE MOVEMENT – GROUP DECIDED TO ADD EVEN MORE WEIGHTS.

$1.00 GLUE STICK
BALANCE OF $35.00

RAN TEST #4 – TIME 39 SECONDS:
TOO MUCH WEIGHT – DECREASED MOVEMENT – ITEM WAS TOO TOP HEAVY – DECIDED TO REMOVE ADDITIONAL WEIGHTS.
GROUP DETERMINED MODIFICATIONS ON DEVICE WERE COMPLETED.
RAN TRIAL RUN #5 — DEVICE MOVED 65 SECONDS.
RAN TRIAL RUN #6 — HELD HANDLE TIGHTER – 67 SECONDS
RAN TRIAL #7 – WENT 68 SECONDS –
- WE ARE READY FOR TOMORROW'S CLASS COMPETITION!
(BALANCE OF $35.00)

Sample Graph for Constant Movement

The problem was to design and build a device that would move the longest period of time.

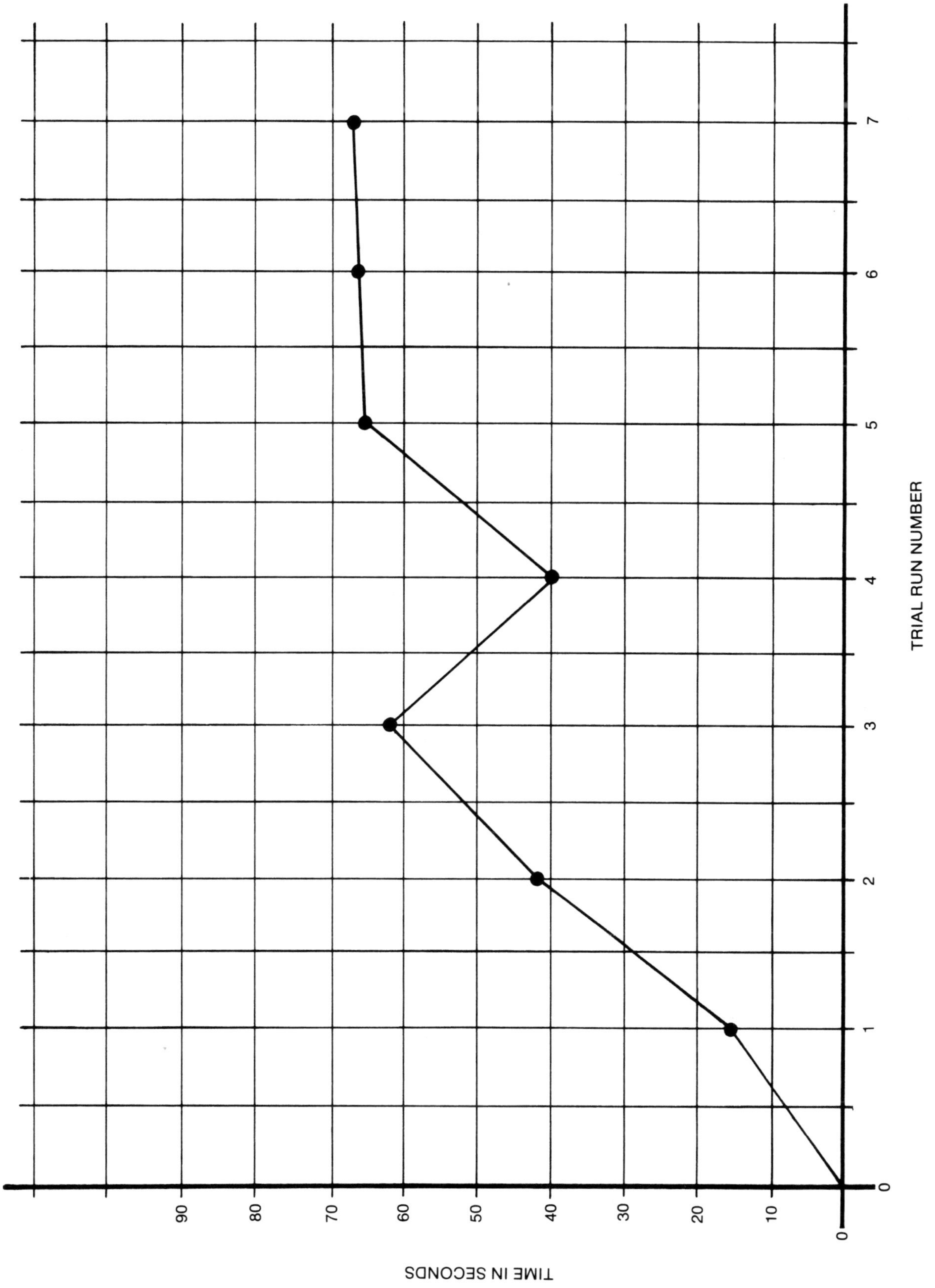

$1.00

TECHNOLOGY MONEY

GROUP NO. ____________

PATENT NO. ____________

ONE DOLLAR

$1.00

$5.00

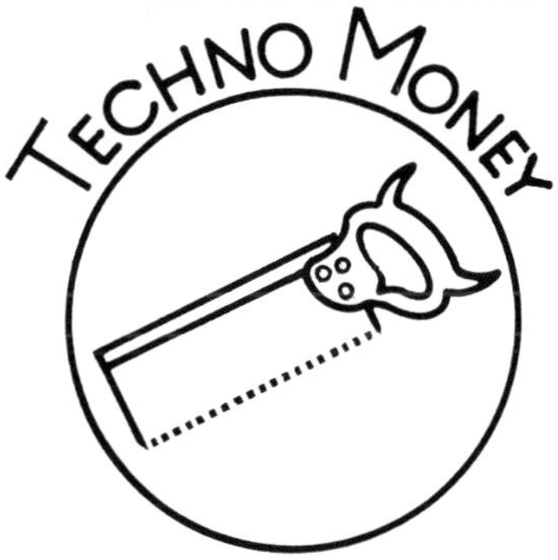

TECHNOLOGY MONEY

GROUP NO. ____________

PATENT NO. ____________

FIVE DOLLARS

$5.00

$10.00

TECHNOLOGY MONEY

GROUP NO. ____________

PATENT NO. ____________

TEN DOLLARS

$10.00

$20.00

TECHNOLOGY MONEY

GROUP NO. ____________

PATENT NO. ____________

TWENTY DOLLARS

$20.00

TECHNO CHARGE

GROUP NO. ____________

GROUP MEMBERS ____________

____

____

PATENT NO. ____________

TECHNO CHARGE

GROUP NO. ____________

GROUP MEMBERS ____________

____

____

PATENT NO. ____________

TECHNO CHARGE

GROUP NO. ____________

GROUP MEMBERS ____________

____

____

PATENT NO. ____________

TECHNO CHARGE

GROUP NO. ____________

GROUP MEMBERS ____________

____

____

PATENT NO. ____________

TECHNO CHARGE

GROUP NO. ____________

GROUP MEMBERS ____________

____

____

PATENT NO. ____________

TECHNO CHARGE

GROUP NO. ____________

GROUP MEMBERS ____________

____

____

PATENT NO. ____________

TECHNO CHARGE

GROUP NO. ____________

GROUP MEMBERS ____________

____

____

PATENT NO. ____________

TECHNO CHARGE

GROUP NO. ____________

GROUP MEMBERS ____________

____

____

PATENT NO. ____________

SAVINGS DEPOSIT

GROUP NO. ________

DATE ________

TOTAL AMOUNT

$ ________

DEPOSIT FOR THE ACCOUNT OF ________ (GROUP NO.)

GROUP MEMBERS ________

____

____

PATENT NO. ________

DATE ________

TECHNO CASH ________

____

____

TOTAL DEPOSIT $ ________

SAVINGS DEPOSIT

GROUP NO. ________

DATE ________

TOTAL AMOUNT

$ ________

DEPOSIT FOR THE ACCOUNT OF ________ (GROUP NO.)

GROUP MEMBERS ________

____

____

PATENT NO. ________

DATE ________

TECHNO CASH ________

____

____

TOTAL DEPOSIT $ ________

SAVINGS DEPOSIT

GROUP NO. ________

DATE ________

TOTAL AMOUNT

$ ________

DEPOSIT FOR THE ACCOUNT OF ________ (GROUP NO.)

GROUP MEMBERS ________

____

____

PATENT NO. ________

DATE ________

TECHNO CASH ________

____

____

TOTAL DEPOSIT $ ________

SAVINGS DEPOSIT

GROUP NO. ________

DATE ________

TOTAL AMOUNT

$ ________

DEPOSIT FOR THE ACCOUNT OF ________ (GROUP NO.)

GROUP MEMBERS ________

____

____

PATENT NO. ________

DATE ________

TECHNO CASH ________

____

____

TOTAL DEPOSIT $ ________

WITHDRAWAL

GROUP NO. ________

DATE ____________

AMOUNT $________

SAVINGS WITHDRAWAL

GROUP NO. ________

PAY THE SUM OF $____________

____________________________ DOLLARS

DATE ________

TO: GROUP MEMBERS ____________,

____________, ____________

WITHDRAWAL

GROUP NO. ________

DATE ____________

AMOUNT $________

SAVINGS WITHDRAWAL

GROUP NO. ________

PAY THE SUM OF $____________

____________________________ DOLLARS

DATE ________

TO: GROUP MEMBERS ____________,

____________, ____________

WITHDRAWAL

GROUP NO. ________

DATE ____________

AMOUNT $________

SAVINGS WITHDRAWAL

GROUP NO. ________

PAY THE SUM OF $____________

____________________________ DOLLARS

DATE ________

TO: GROUP MEMBERS ____________,

____________, ____________

WITHDRAWAL

GROUP NO. ________

DATE ____________

AMOUNT $________

SAVINGS WITHDRAWAL

GROUP NO. ________

PAY THE SUM OF $____________

____________________________ DOLLARS

DATE ________

TO: GROUP MEMBERS ____________,

____________, ____________

Patent Application

THIS APPLICATION FOR PATENT RIGHTS IS PRESENTED BY ____________________ .

GROUP MEMBERS ARE____________________, ____________________, ____________________.

THIS PATENT GIVES THE ABOVE NAMED, THE RIGHTS FOR THE FOLLOWING DEVICE:

(SKETCH GROUP'S METHOD OF SOLUTION)

UPON APPROVAL OF THIS APPLICATION, WE HAVE RECEIVED THE FOLLOWING PATENT NUMBER:____________________. A FEE OF $5.00 (PER GROUP MEMBER) IS ENCLOSED WITH THIS APPLICATION.

Name	Group Members
Selected Item	

Activity III-1
Brainstorming

Your instructor will guide the class through a brainstorming process. The businessman below is ready for a very important meeting, which begins in 10 minutes. The meeting is to be held on the tenth floor of the building he is standing in front of.

It is a super hot August day. The man just purchased an ice cream cone. Due to the heat, the ice cream fell off the cone onto his shoe and instantly it began to melt. What can he do? Remember, in 10 minutes he must be on the tenth floor for a very important meeting.

With your instructor acting as the recorder, the class will run through a brainstorming session to help determine a solution.

Copy the responses below:

____

____

____

____

____

Name	Group Members
Selected Item	

Activity III-2
The Problem and Its Parameters

Your instructor will state a problem that you and your group members will have to solve. The problem could be one of the following, or your instructor could give one. In either case, your group will be given a bag with various resources in it. Your group will be in competition with the other groups in the class to see whose device solves the problem the best. Below are the problems:

1. FLYING DEVICE. Design and build a device to go the highest distance. You must be able to state the height reached. You may use one outside fastening method, the bag of parts, drafting and design supplies, tools and machines rented with technology money, one outside material, and any amount of cloth desired. A daily log and budget must be kept.
2. DO-NOTHING MACHINE. You are to design and build a do-nothing machine. This machine must have constant movement for the longest possible time. You may start the device, then it must move on its own. Use only the materials in the supply bag, one outside adhesive, and you must use at least 2, but not more than 6, outside resources of the same type. Resources purchased with the technology money may also be used. A daily log and budget must be kept.
3. PROTECTIVE BOX. Design and build a box to protect the fragile item. You are allowed to use only one piece of 18″ × 24″ cardstock, scissors, silkscreen materials, and drafting tools. Box must be attractive with a silk screen logo on it. One-color paint and one-color ink may be used. You may use only the tools and machines rented with the technology money. A log and budget must be kept. Your instructor will have the device you must protect.
4. LEARNING GAME. Using any subject (math, science, music, for example) your group will design and build a game that will help one learn about that subject. You may use resources in the supply bag and those purchased with the technology money. You may use the lithographic and stencil printing areas. Drafting and design tools may be used along with 2, but not more than 6, outside materials. The game board or part of it must be printed with at least one of the above printing methods. Keep log and budget.
5. BANK. Using the materials in the supply bag plus the resources purchased with the Techno Money, design and build a bank that will sort and count the coins placed in it. Keep a daily log and budget.
6. FIRE-OUT DEVICE. Design and build a device that will put out a candle 3′ away. You may use the materials in the supply bag plus the resource purchased with the technology money. The device *may not* tip the candle over. A log and budget must be kept.
7. BALLOON DEVICE. You are to design and build a balloon-powered device that will travel the farthest distance on fixed suspended string. The device must be propelled by the balloon. The device must be easily set on the string. You may use the materials in the supply bag and those purchased with the Techno Money. A log and budget must be kept.

WHAT YOU ARE TO DO:

1. Complete problem solving introduction sheet (by either teacher's lesson or reading completed) on the next page.
2. Working in groups of three (select two other classmates with whom you have not worked), begin brainstorming. Your instructor will give your group a group number.
3. Keep a daily log and budget. (Proper planning may save money.)
4. Using the provided sheets and materials, start your problem solving activity.
5. After your device is completed, compare your solution to those from other groups.

Problem Solving Introduction Sheet

What is brainstorming? ______________________________

____

____

____

List the seven steps of problem solving:

1. ______________________________
2. ______________________________
3. ______________________________
4. ______________________________
5. ______________________________
6. ______________________________
7. ______________________________

What is a patent? ______________________________

____

What is interest? ______________________________

____

What are the two methods where interest may be applied? ______________________________

____

____

____

What is a budget? ______________________________

____

____

Technology Log

Name ______________________

Group Members ______________________

____

____

Group Number ______________________

Patent Number ______________________

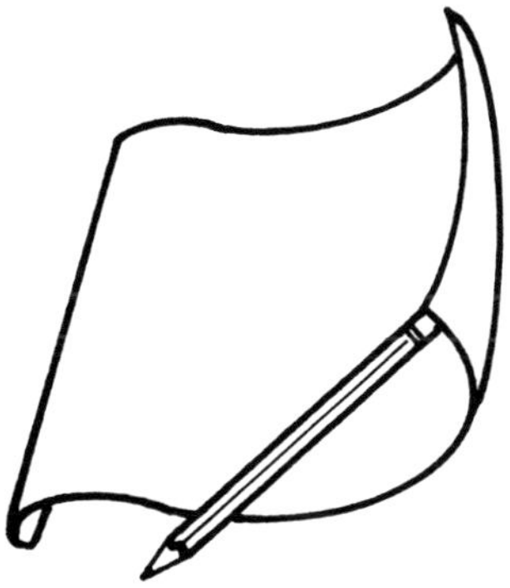

Contents

PROBLEM STATEMENT

GROUP'S BRAINSTORMING IDEAS

DEFINE THE PROBLEM

PARAMETERS

Possible Solutions

(sketches developed from group)

Final Solution

Sketch group's selected solution.

Log Keeping

Log Keeping

Log Keeping

Log Keeping

Modifications

LOG EVALUATION

Date ______________________ Remarks ______________________

Date ______________________ Remarks ______________________

Final Evaluation ______________________ Remarks ______________________

UNIT IV

Systems

We are surrounded by examples of technological systems, which are in place to help people cope more easily with their environment. These systems are designed and built to meet specific human needs and wants.

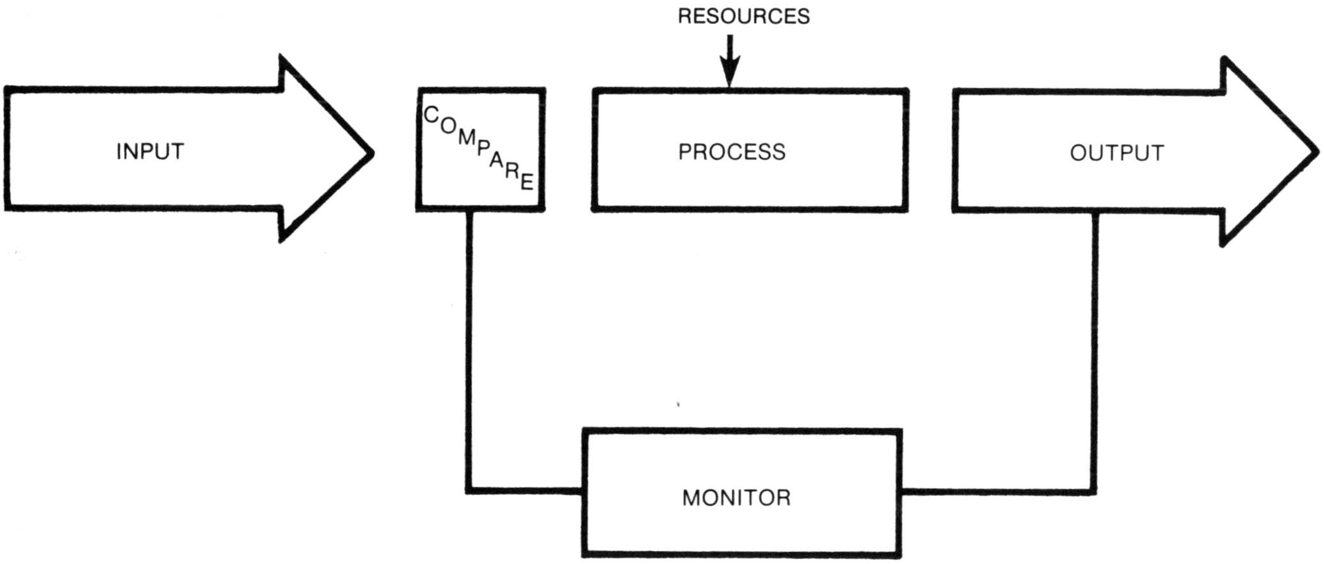

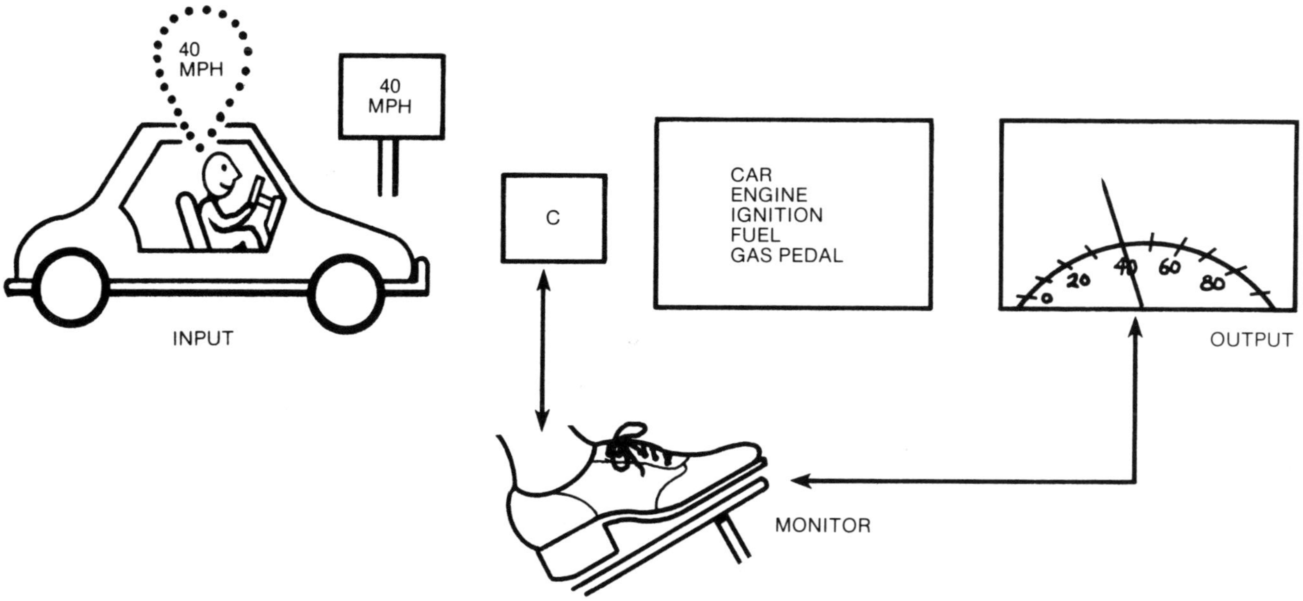

These systems can be built with a variety of parts or methods. In other words, the systems may be referred to as one or a combination of the following: natural, mechanical, hydraulic, pneumatic, and electrical control operated systems.

The solar system is an example of a natural system over which we have no control.

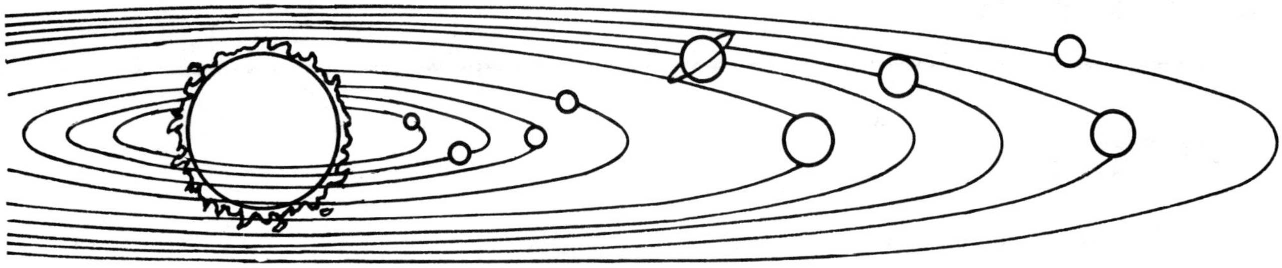

The Solar System

The mechanically operated system uses mechanical devices to operate. Some of the more popular mechanical devices are the cam, gear, lever, chain and sprocket, and the pulley and belt.

The *cam* is a rotating device that has a high spot (lobe) on one side. As the cam rotates, the lobe will push on a follower, causing it to rise. The cam is in the small gasoline and the automobile engine.

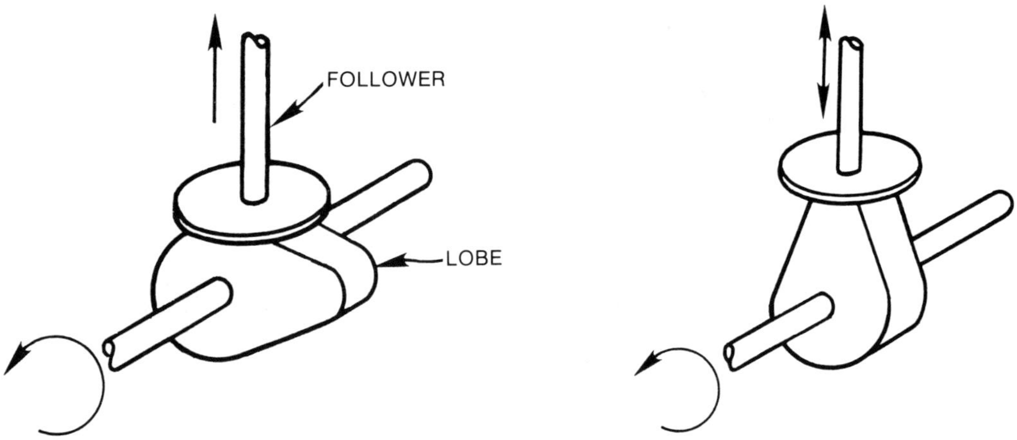

A Cam

The *gear* is like two wheels riding on each other, one causing the other to rotate.

Gears are used in many devices in the world today. The wind-up watch is a familiar example. Enclosed in these watches are many gears to do a specific task.

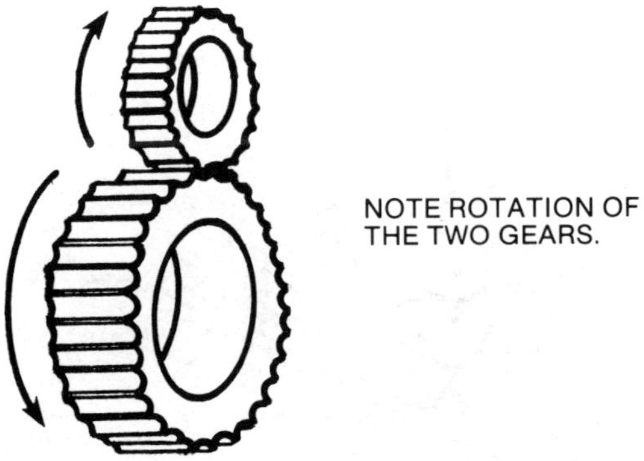

Gears

Levers are usually used in conjunction with other mechanical devices to help a system operate. The picture below shows a lever being operated by a cam.

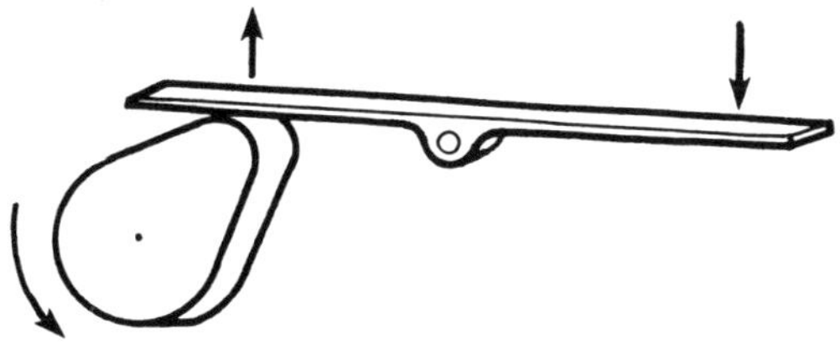

Lever Operated by a Cam

The *chain and sprocket* are mechanical devices that many of you have used. The chain and sprocket system is the mechanical device that makes your bicycle move when you push on the pedals.

Chain and Sprocket

The *pulley and belt* are mechanical devices much like the chain and sprocket, except that the pulley and belt operate by tension and friction between them. The cassette tape deck in your stereo or radio is driven by a pulley and belt.

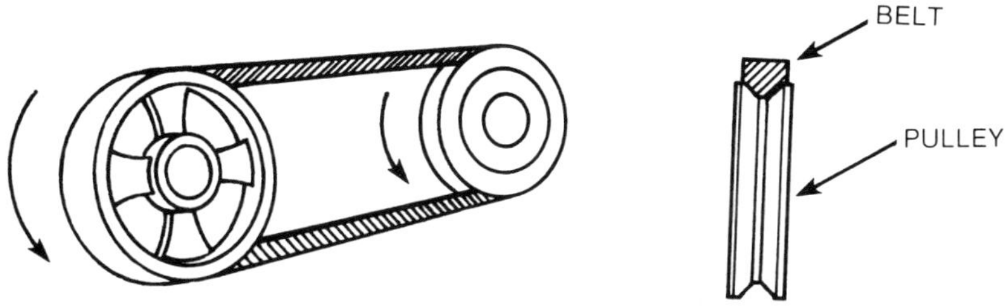

Pulley and Belt

Hydraulic operated systems are ones in which the power is transmitted by the use of a fluid (such as oil). Fluid traveling from one chamber to another can do a specific task.

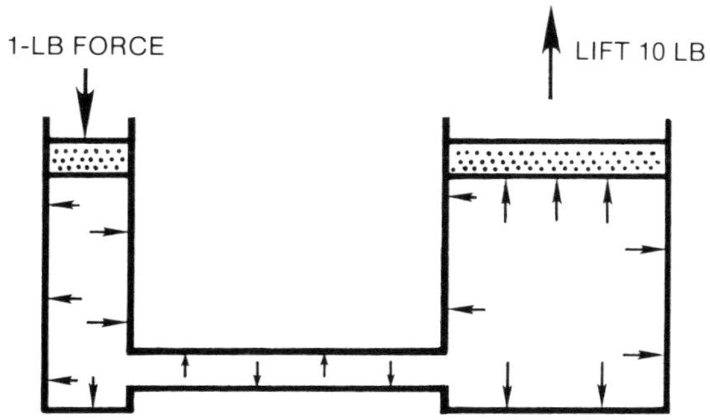

Hydraulic System

A good example of a hydraulic operated system is the brake system in most automobiles today.

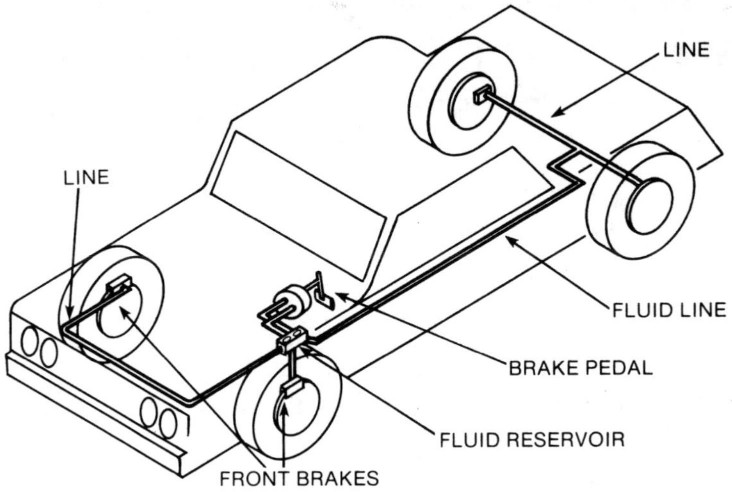

Brake System

Pneumatic devices are similar to the hydraulic systems. The exception is that hydraulic machines use fluid to transmit the power, where as pneumatic machines use a gas (such as air) to transmit the power. An example of a pneumatic tool is the tire pump.

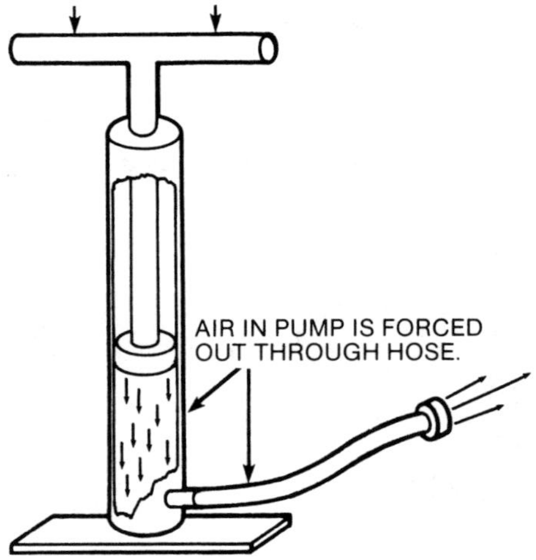

Electrical operated devices are used for many purposes throughout the world, such as alarm clocks, conventional and microwave ovens, televisions, and computers. These devices were designed and built to meet a specific need and want.

Today, many devices use electricity to the specific task. The electricity is controlled by electrical components that are designed and built for a desired result in the operation of a machine.

A schematic drawing shows the designer the path that electrons are to travel. Schematic symbols are used to represent the electrical components. This illustration shows a schematic drawing of the standard flashlight.

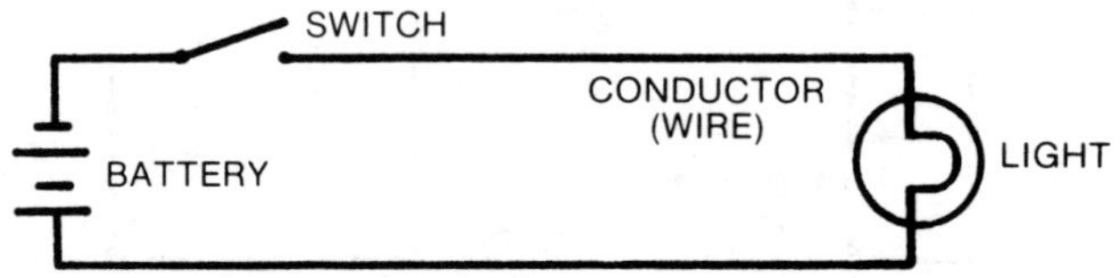

Schematic Drawing of a Flashlight

The schematic drawing is then converted to the mechanical drawing. The mechanical drawing shows how the various parts would appear if the device were disassembled. Here is a mechanical drawing of the flashlight.

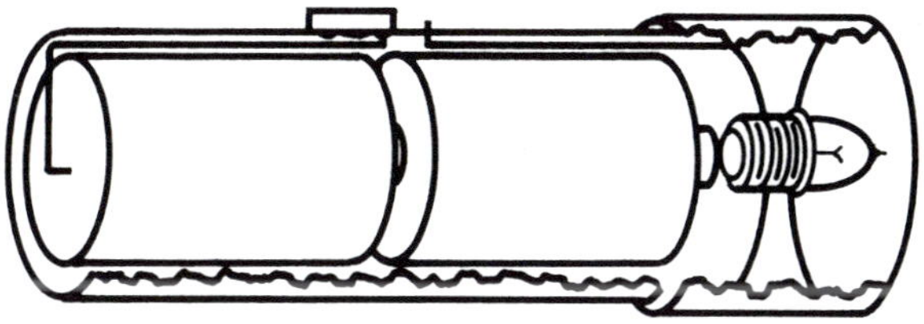

Mechanical Drawing of a Flashlight

SYSTEMS AND SUBSYSTEMS

The device (system) designed and built to meet a human need and want may utilize any of the methods above or by a combination of these methods. Each series of parts that are to perform a specific task are referred to as a system. These systems are all combined to make a larger system—meaning that they now become a *subsystem* of the larger system. All of the subsystems work together in such a way that the device built will meet the desired need or want.

For example, let's look at the bicycle as a system. The bicycle was designed to meet a human need or want. It too has subsystems that are in place to make the bike meet our needs.

Bicycle

Think of the systems around us and what subsystems they comprise. We have just learned some of the most common methods used in the operation of devices. Many of these tools and machines are frequently controlled and adjusted automatically to changing conditions. On the other hand, some machines may require human (manual) control.

The performance of any device is controlled by one of two methods: open loop system or closed loop system.

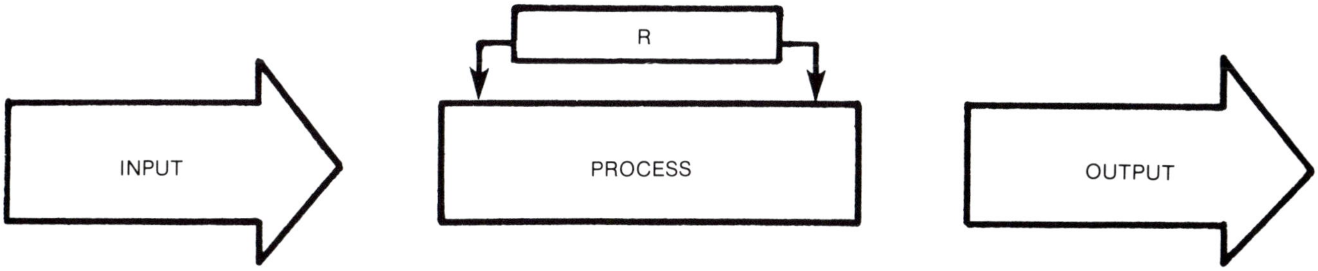

The Open Loop System

Look at the systems model at the bottom of the previous page:

Input = What is to be done

Process = What has to be done
to accomplish our task

Output = The end result

R = The resources of technology

An example of an open loop system is shown below.

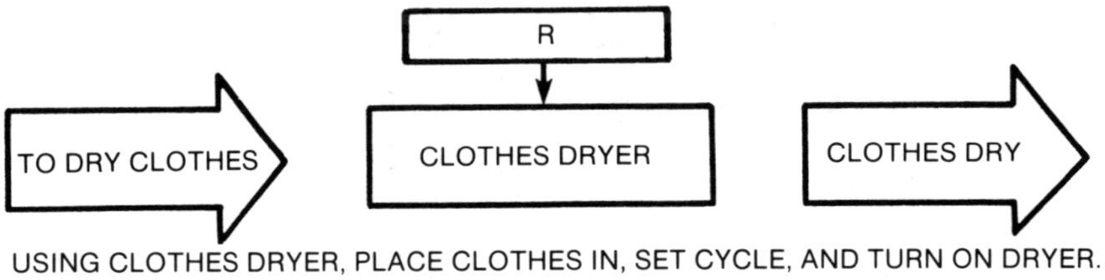

Clothes Dryer as an Open Loop System

A closed loop system is shown in this diagram.

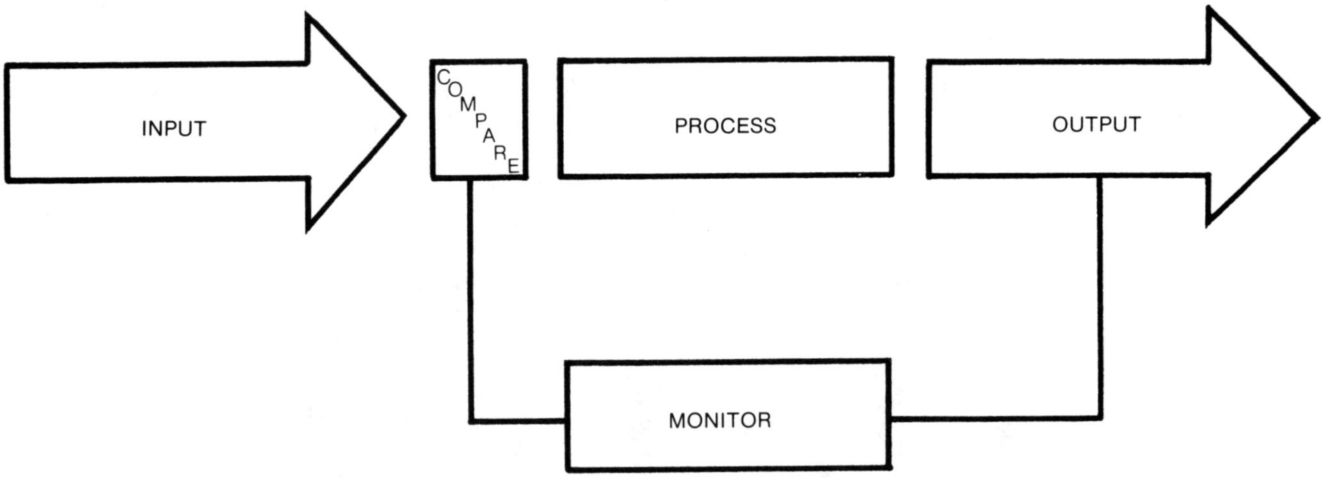

Closed Loop System

Again, there is a desired task, the input. The resources are used to perform a process that will give an output. Unlike the open loop, this system has a monitor that will compare the input to the output. If the desired input is not met, the monitor will turn on the process again until the desired output is accomplished.

The monitor is what is referred to as the feedback part of the system. Remember the feedback will compare the output to the input, and if they are not the same, the process will keep operating until the output meets the input.

Let's look at a typical everyday device to see if it is an open loop or a closed loop system, for example, the clothes dryer.

Wet clothes are placed in the machine. The dry cycle is set and the machine begins its task. When the dryer has completed its cycle, it goes off. Are the clothes dry? Somebody must open the dryer and check them. If the clothes are still wet, the cycle is turned on again to meet our input. Up until the clothes are checked, the dryer is an open loop system.

If sensors to feel moisture were placed in the dryer, the sensors would automatically turn on the dryer and the dryer would continue the process until the clothes were dry. Thus, the dryer is a closed loop system, as shown at the top of the next page.

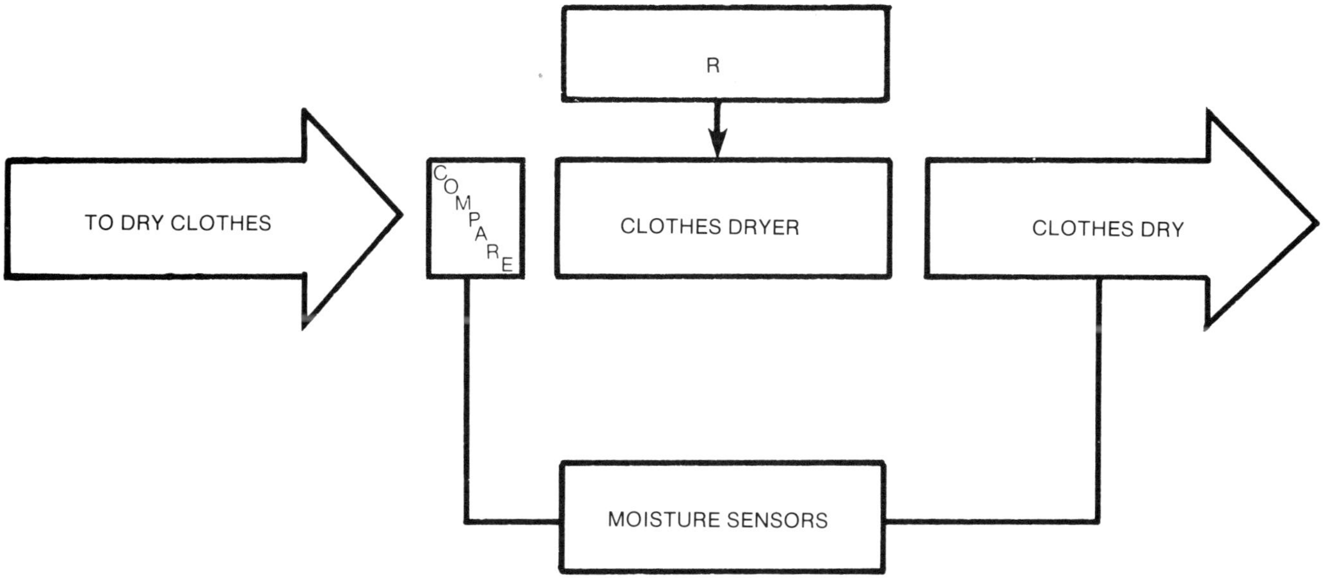

Clothes Dryer as a Closed Loop System

Computers Are Used in Many Places to Control and Operate a System.

Name	Group Members
Selected Item	

Activity IV-1
Identifying Controls

Identify what type of control device the items below are:

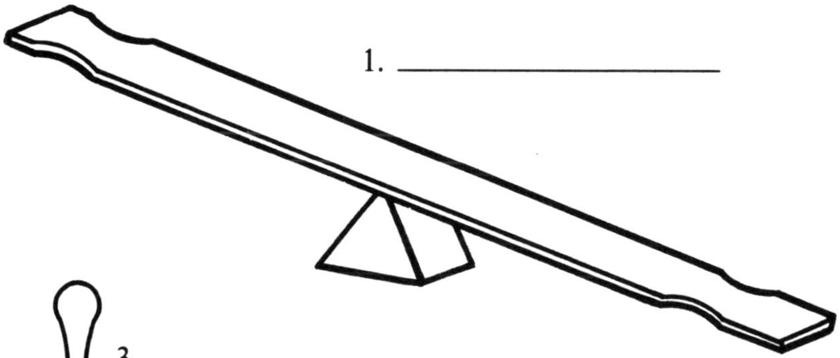

1. ____________________

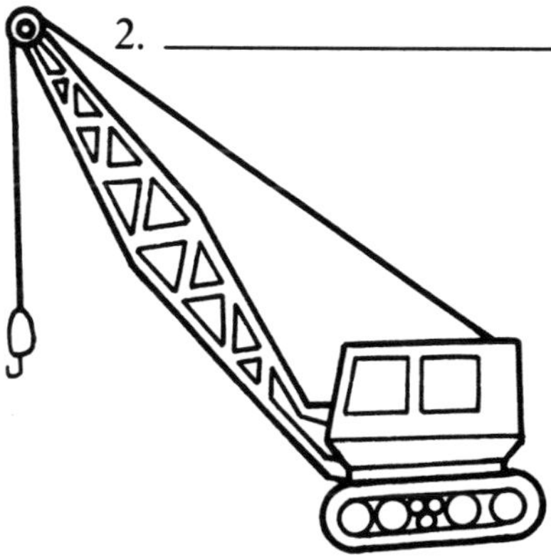

2. ____________________

3. ____________________

4. ____________________

5. ____________________

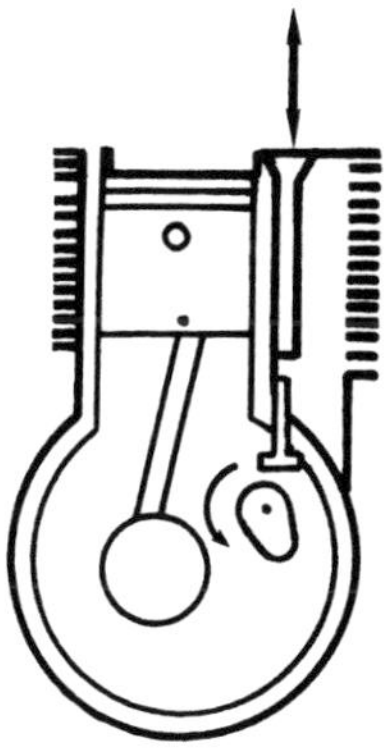

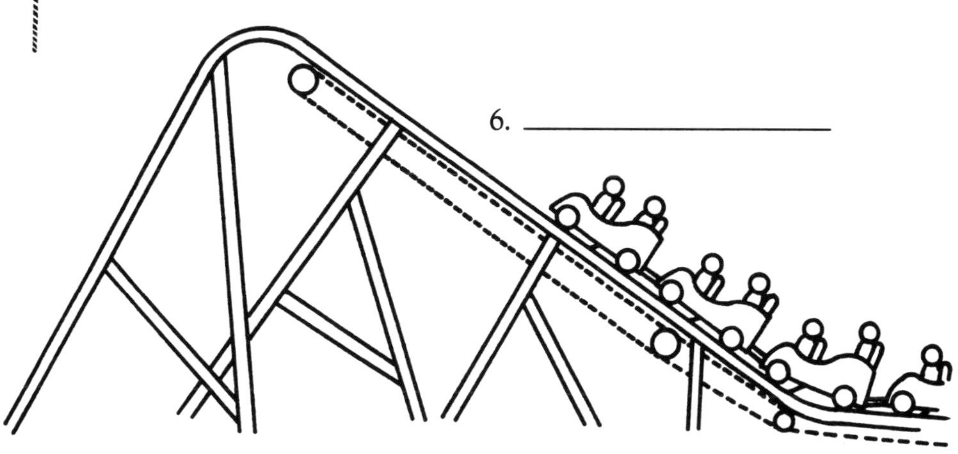

6. ____________________

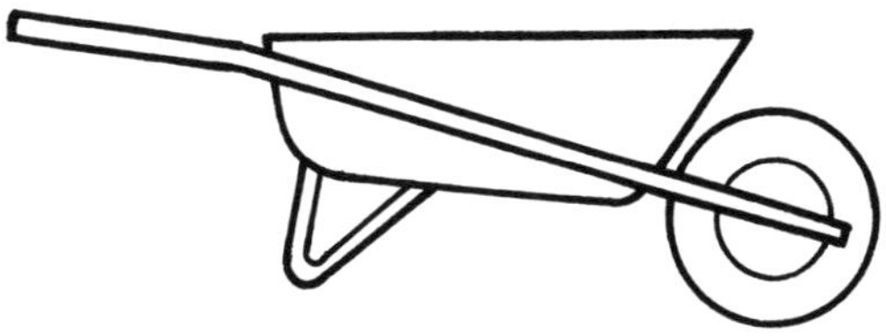

7. ____________________

Name	Group Members
Selected Item	

Activity IV-2
Systems and Subsystems

With your instructor leading the class, you are to list how the two living systems differ due to the subsystems they have. (List what the present day house has that the house of 1800s did not.)

Living Systems for Humans in 1800

List its subsystems:

____

____

____

____

____

____

Living Systems for Humans at Present

List its subsystems:

____

____

____

____

____

____

SYSTEMS/SUBSYSTEMS

List how the two systems differ due to their subsystems.

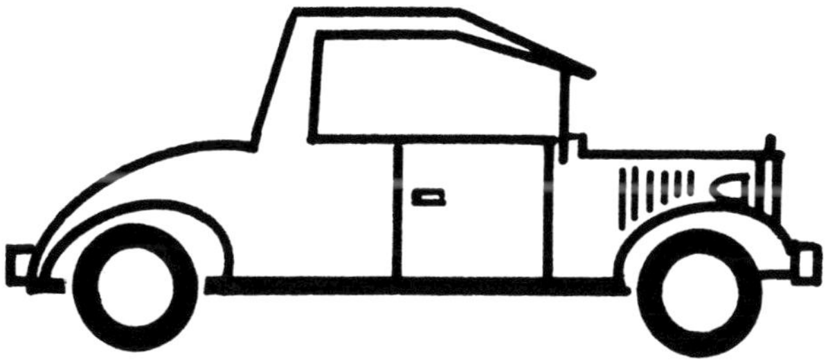

Human Transport System in 1920

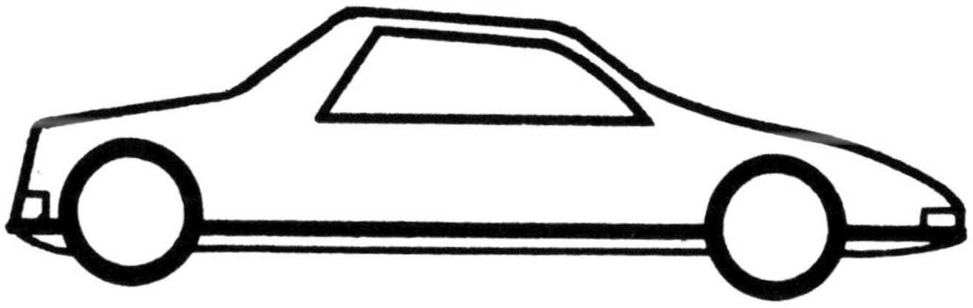

Human Transport System at Present

Name	Group Members
Selected Item	

Activity IV-3
Open Loop Systems

Identify four items (devices) in the world that you have either seen or used that are open loop systems. Set each of these in the open loop diagrams below.

1. ______________________

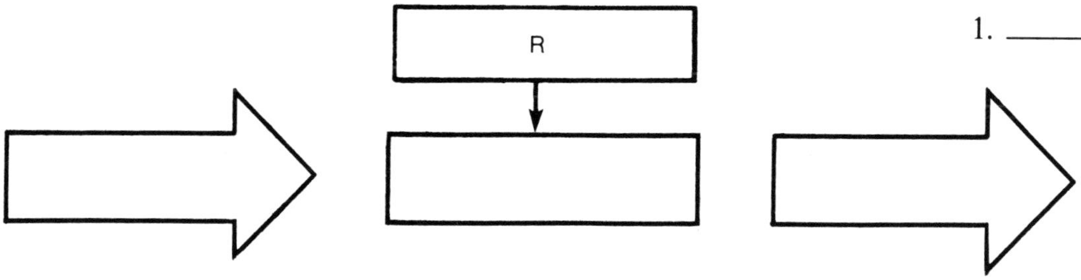

2. ______________________

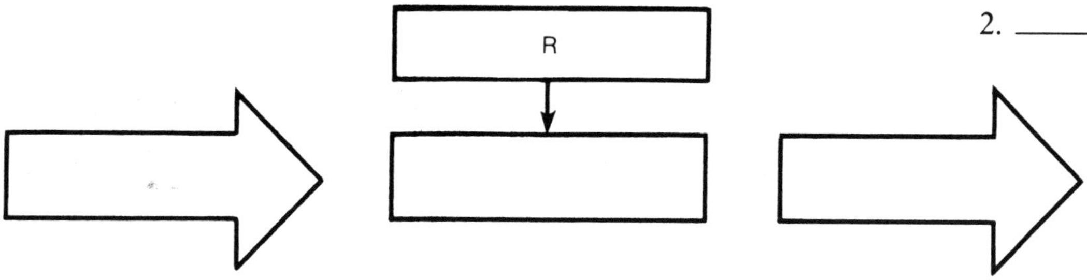

3. ______________________

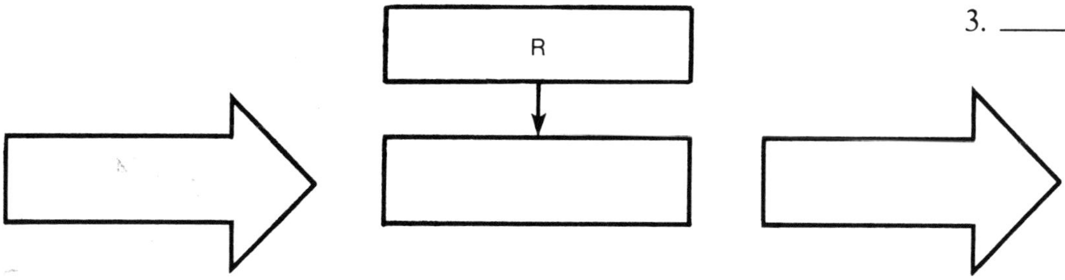

4. ______________________

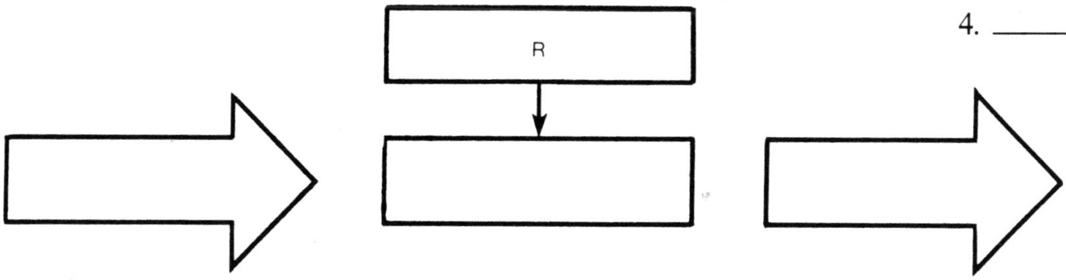

Name	Group Members
Selected Item	

Activity IV-4
Closed Loop Systems

Identify four items (devices) in the world that you have either seen or used that are closed loop systems. Set each of these in the closed loop diagrams below.

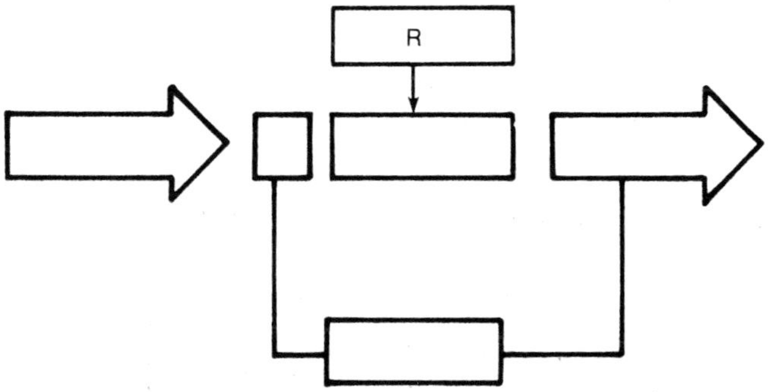

1. ____________________

2. ____________________

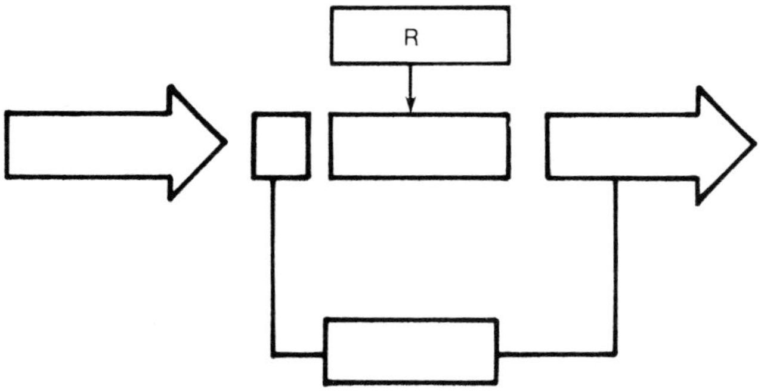

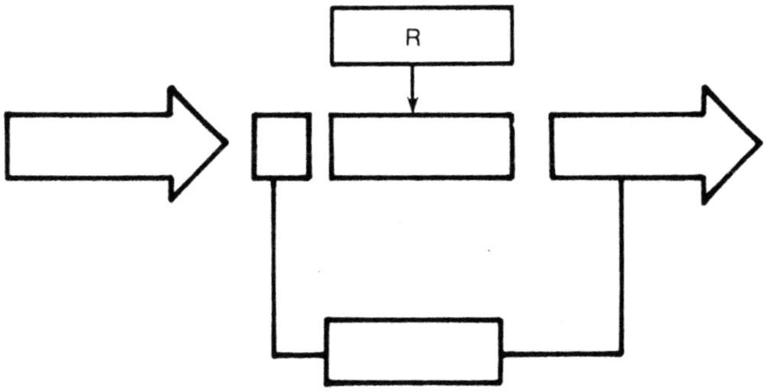

3. ____________________

4. ____________________

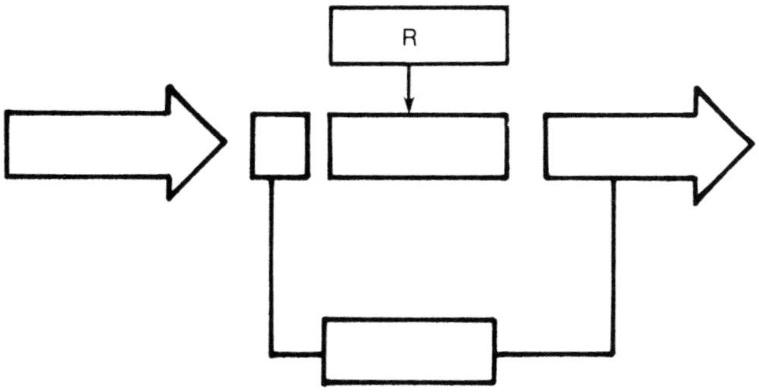

Name	Group Members
Selected Item	

Activity IV-5
Building a System

WHAT YOU ARE TO DO:

Using the various resources, you are to design and build a system to meet a specific task (need or want). Your system must contain at least one subsystem. Show all your designs on the provided paper.

You must also:

1. State system's task: what the system's input is.
2. Sketch where the system is going to be located (for example, water alarm to be placed in basement).
3. Do a detailed drawing of the system's operation (schematic and drawing of the parts needed).
4. Using either a 3-view drawing or a pictorial drawing show the complete system design.
5. Build the system.
6. Diagram your system, indicating whether it is open or closed loop.
7. Demonstrate to the class how your system operates.

You may use one of the system ideas enclosed in this unit or think of your own.

Water-in-the-Basement Alarm

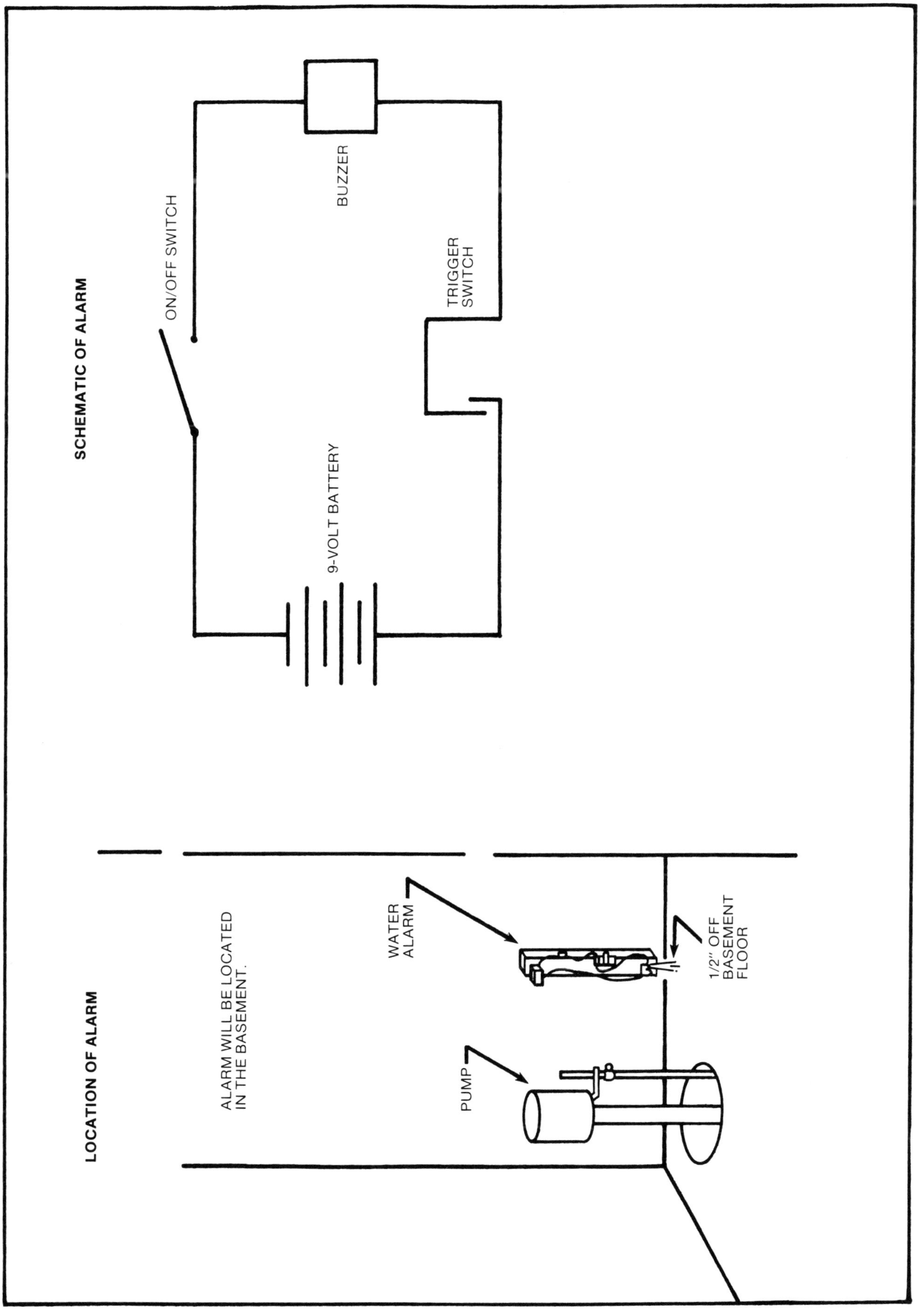

Water Alarm Location and Schematic Drawing

Parts Drawing

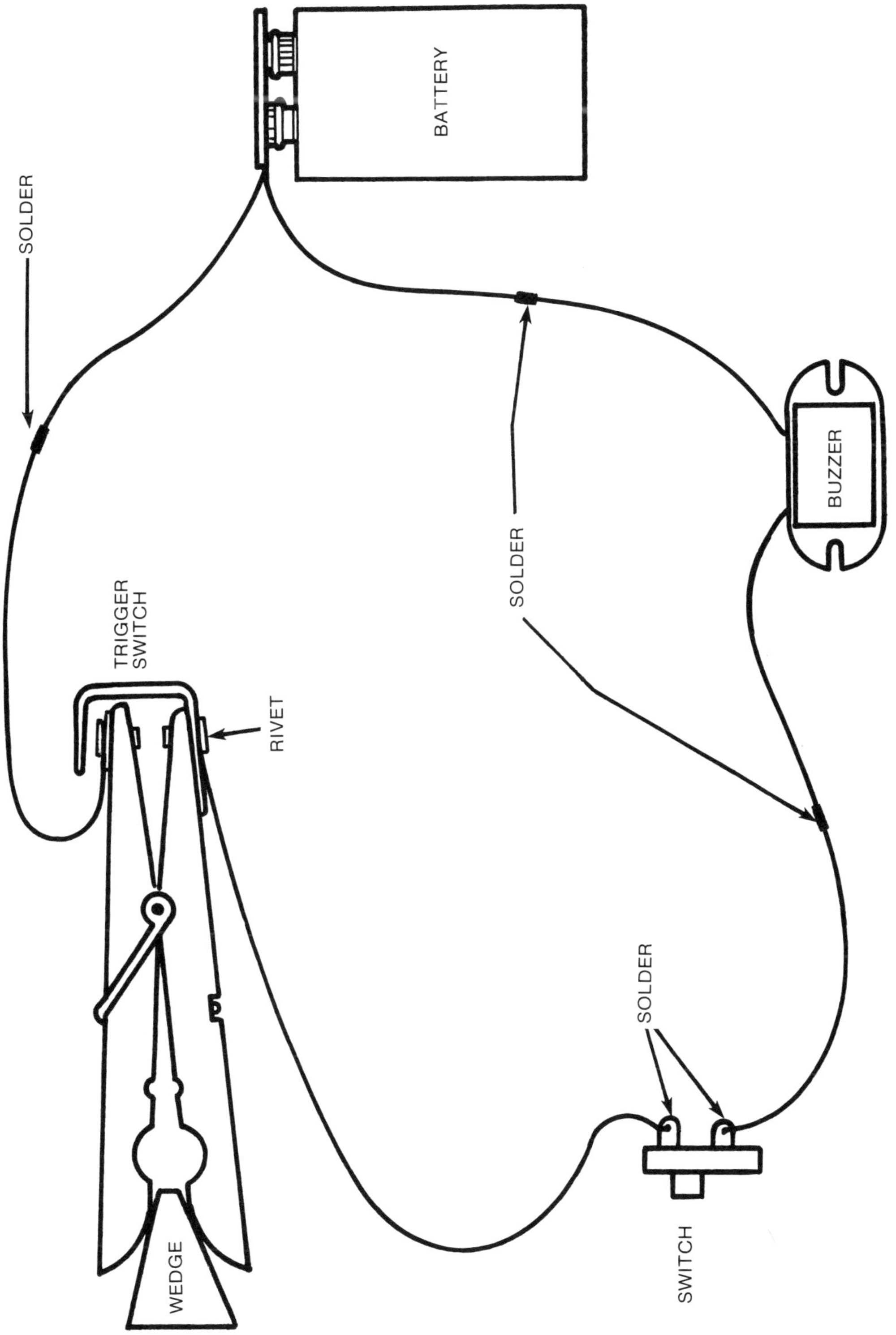

Assembly Drawing

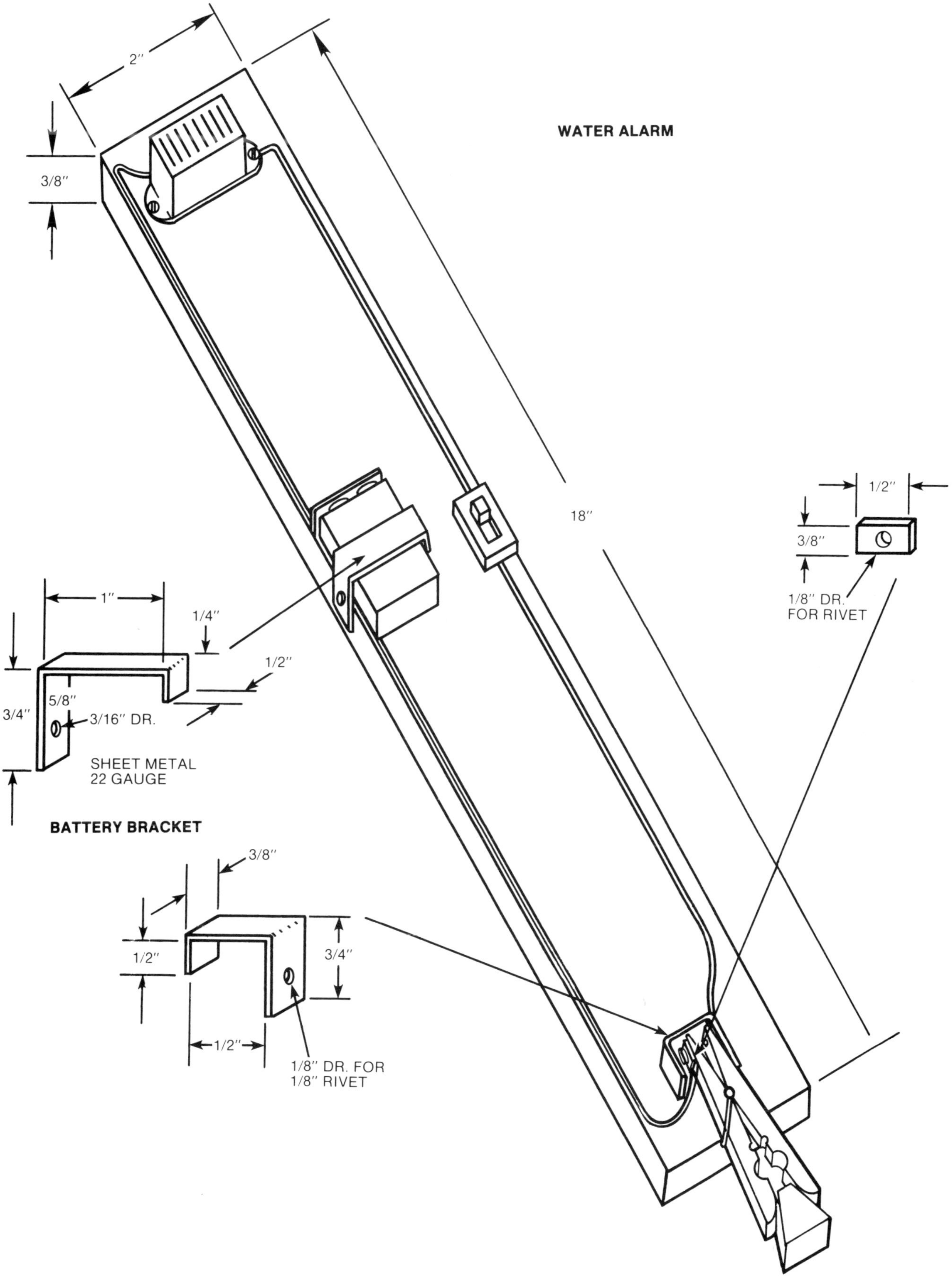

Sketch where your system is going to be located.

Draw your system's schematic.

Design your system's parts on this page.

Use this page for your system's assembly drawing.

Diagram your system here. (See pages 175–177 for diagram format.)

Other System Ideas You May Use

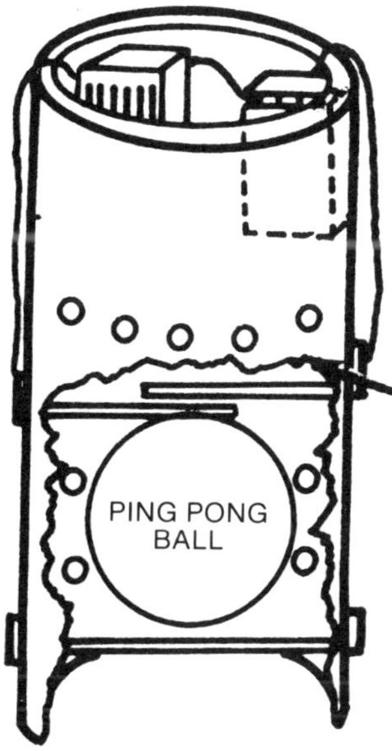

WATER ALARM

- PVC TUBING 1-1/2" DIA.
- PING PONG BALL FLOATS, CAUSING ALARM TO SOUND PUSHING SHEET METAL TO SET ALARM.

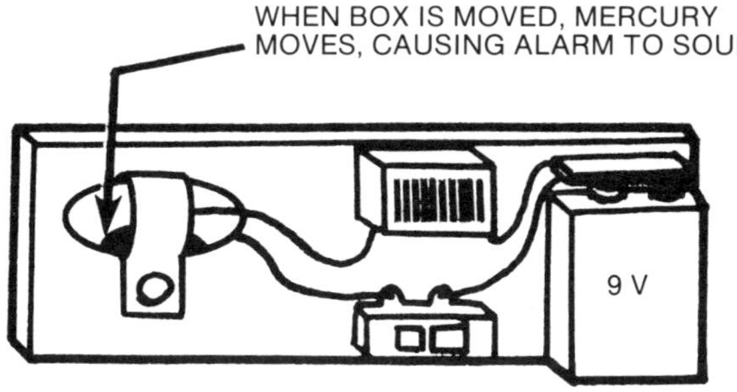

MERCURY SWITCH-MOVEMENT ALARM

FLOOR SENSOR

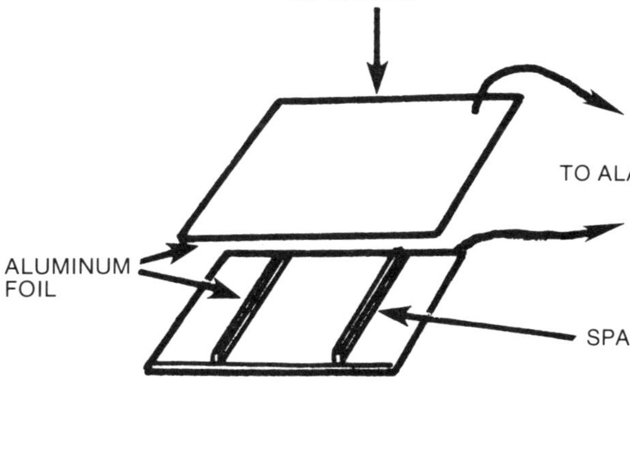

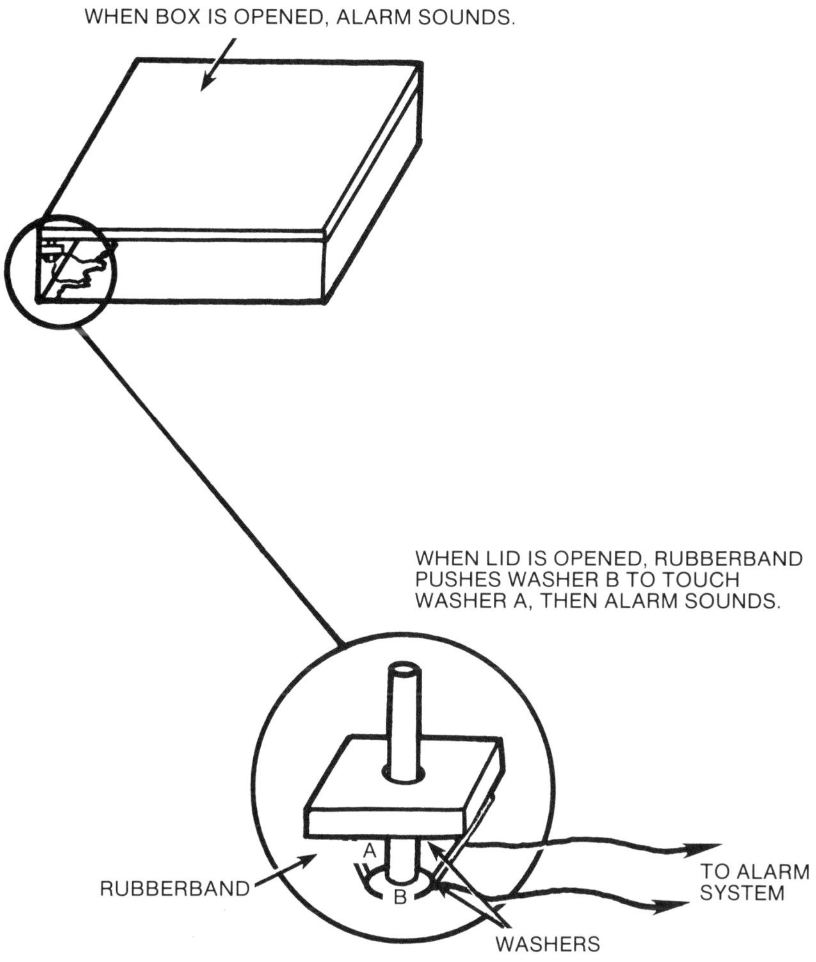
WHEN BOX IS OPENED, ALARM SOUNDS.
WHEN LID IS OPENED, RUBBERBAND PUSHES WASHER B TO TOUCH WASHER A, THEN ALARM SOUNDS.
A
B
RUBBERBAND
TO ALARM SYSTEM
WASHERS

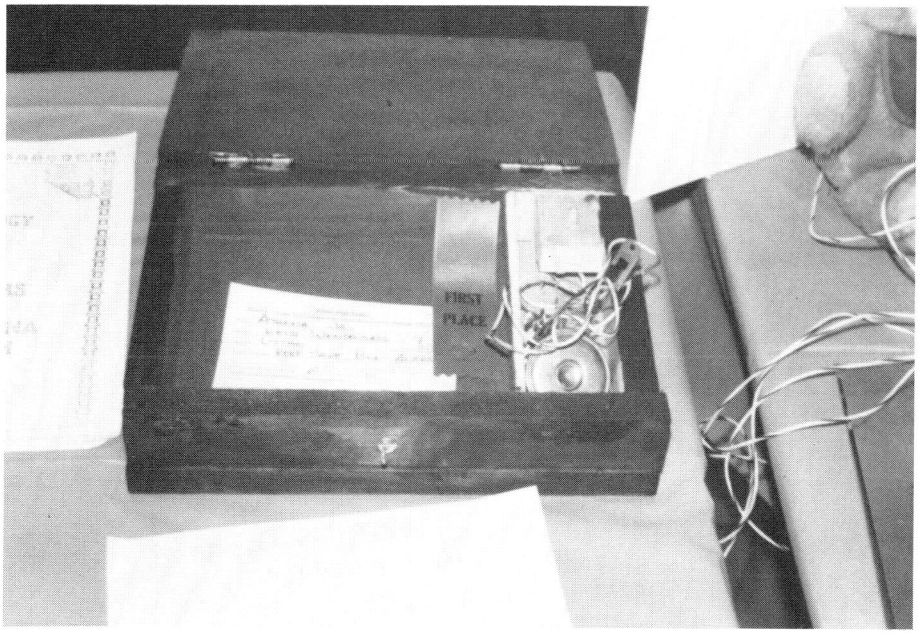
FIRST PLACE

UNIT V

Effects of Technology

We have learned in the past units that technology has changed the world in many ways. A majority of these changes have been accomplished to improve life. Some of these changes could have left unexpected and undesirable results.

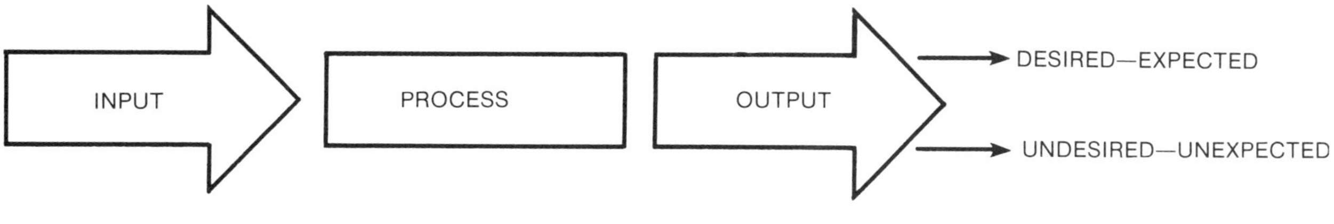

In today's world, biotechnology enables an artificial heart to replace the human heart. The use of artificial limbs move and operate much like the natural limb. Biotechnology provides increased crop yields with the use of hydroponic greenhouses. On the other hand, germ warfare is a possibility that can cause great harm.

Informational technology gave us the prompt communication ability of today. Television, telephone, radio, and the computer aid in the communication process. All are a great plus to life when they are used properly. What happens when these tools are used in a manner they were not intended? We have an unexpected and undesirable result.

Due to physical technology, we have our shelter, clothes, and the ability to transport goods and people. Let's look at our large fifteen to twenty floor skyscrapers used for apartment complexes. What happens to the physically handicapped when he or she must climb stairs? What happens to all, if a fire breaks out? All of these examples have desirable and expected outputs. But, each could have unexpected and undesirable outputs.

Technology must be adapted to the needs and wants of the human. Adapting technology to the human user is what is referred to as *ergonomics.*

Many devices produced are designed and built for the average human. What happens if a person is not average? The device must be adapted to this person. For example, shown below is the inside of an automobile built for the average human. Inside this auto is a very short person. Notice how this person cannot reach the pedals nor see through the windshield.

By adapting technology to the human, the short person can now reach the pedals and see through the windshield. Many times, technology must be adapted to meet the human user.

Before a specific technology development is performed, one must try to evaluate and predict the output. Not just the desired output, but one must try to determine if there are any undesirable outputs.

The use of the system diagram will help predict all possible outputs.

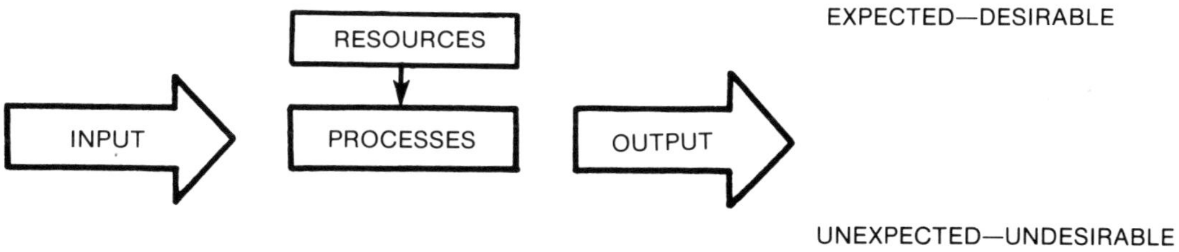

Let's look at space travel and its outputs in the system approach diagram.

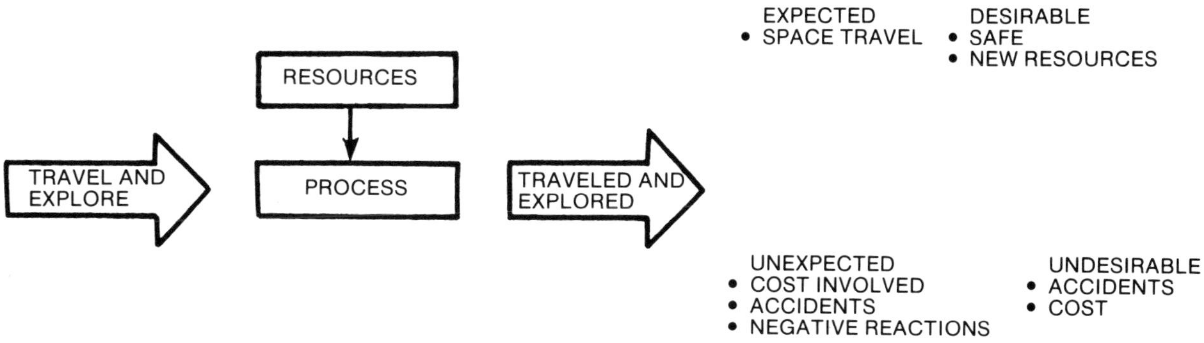

In this unit, you will discover:

1. How technology has been adapted to the human user.
2. The effects of technology, either expected, desirable, undesirable, or unexpected.

Name	Group Members
Selected Item	

Activity V-1
Adapting Technology

WHAT YOU ARE TO DO:

Take each of the following situations and sketch how we have or how we could adapt technology to the human user in each case.

1. In the past, metal tires were used on tractors. The operator found these to be very rough riding.

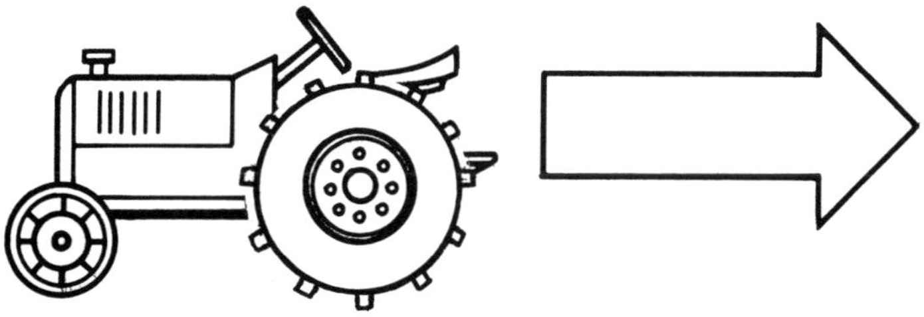

2. People in wheelchairs also have a desire and need to go up the stairs.

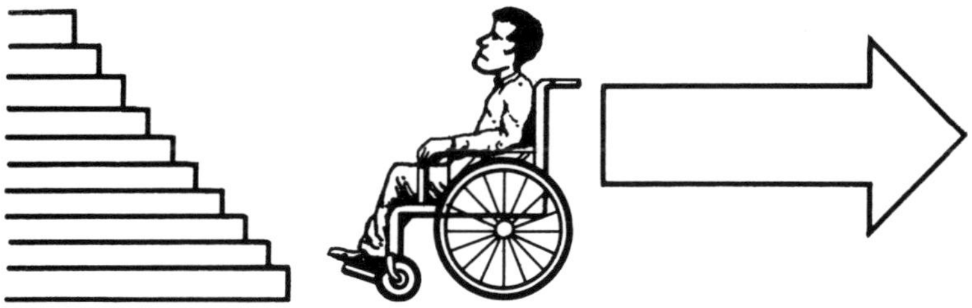

3. The man works odd and long hours. The banks are closed when he gets out of work.

4. A deaf person also has a need and want to use the telephone.

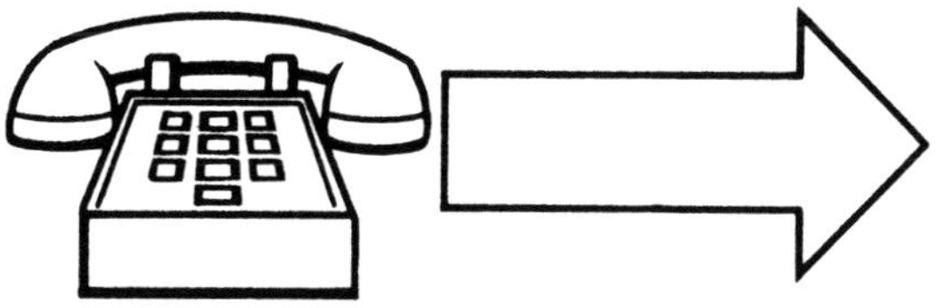

5. The lady's shopping cart is full. She wishes to go out, but she must push the doors open.

6. Take a situation of your own and sketch your adaptation.

Name	Group Members
Selected Item	

Activity V-2
Technological Outputs

WHAT YOU ARE TO DO:

In the system diagrams below, take each of the technological developments listed and state all possible outputs.

1. Nuclear power plant.

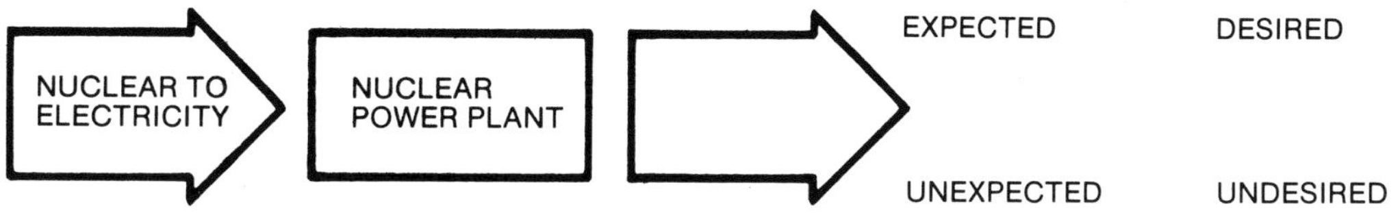

2. Artificial heart.

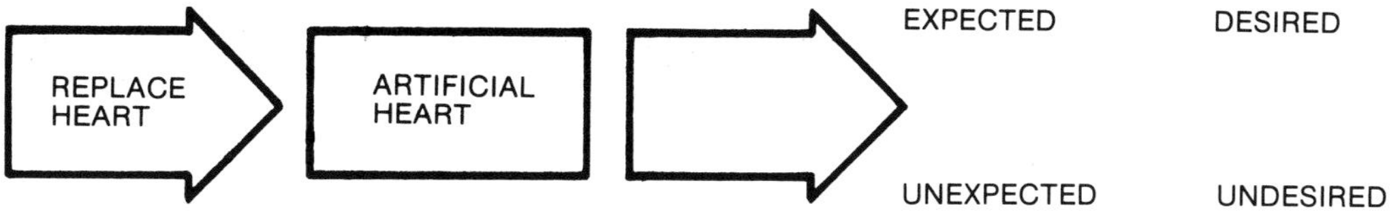

3. Chemical plant.

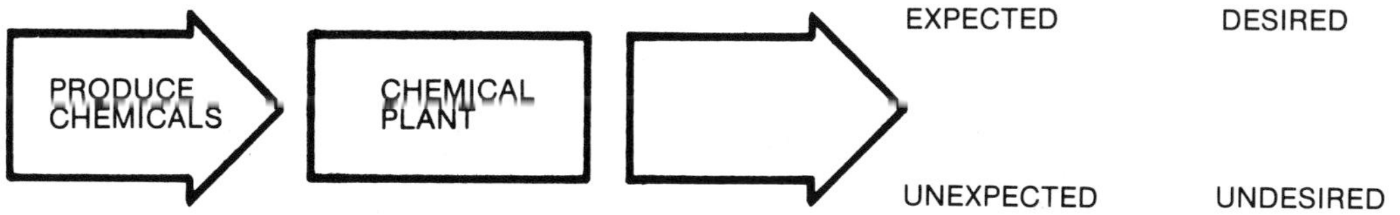

4. The spraying of vegetable plants.

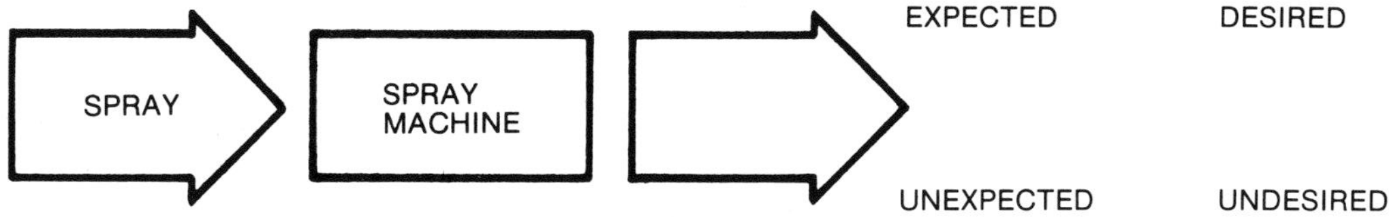

5. Television.

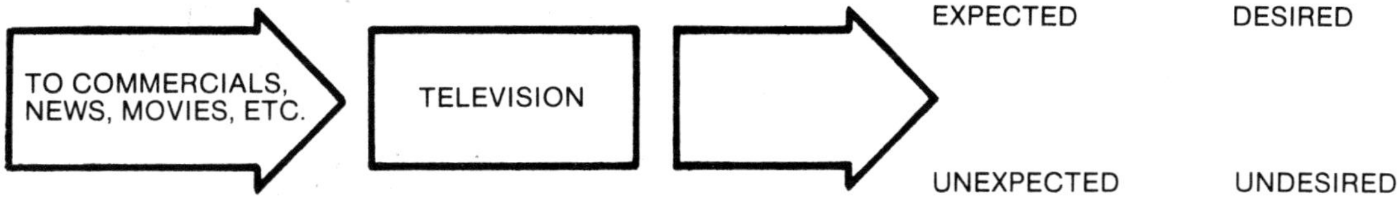

6. Robotic system used in industry to move, sort, and drill.

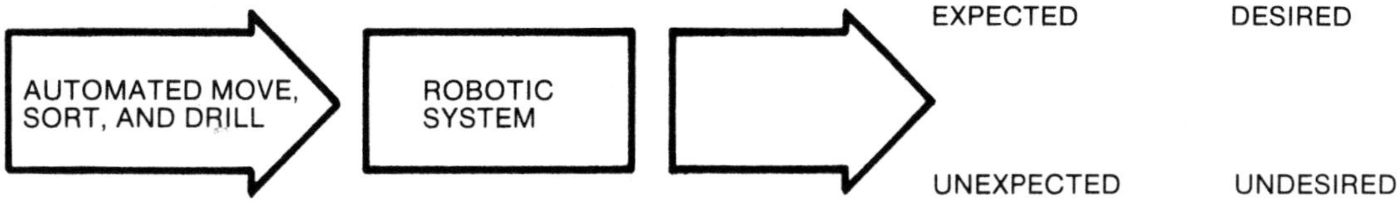

7. Automobile.

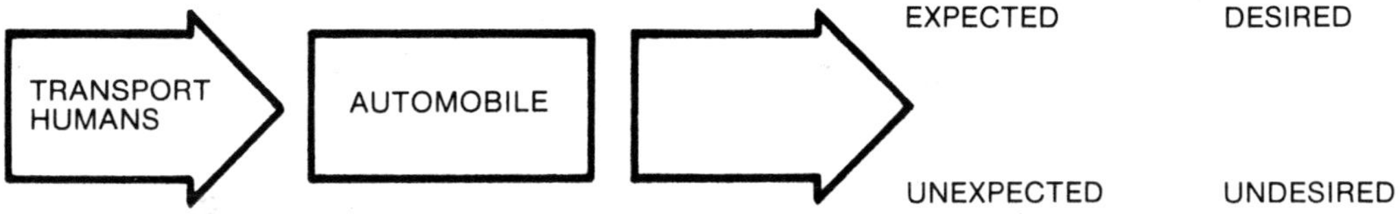

8. Microwave cooking.

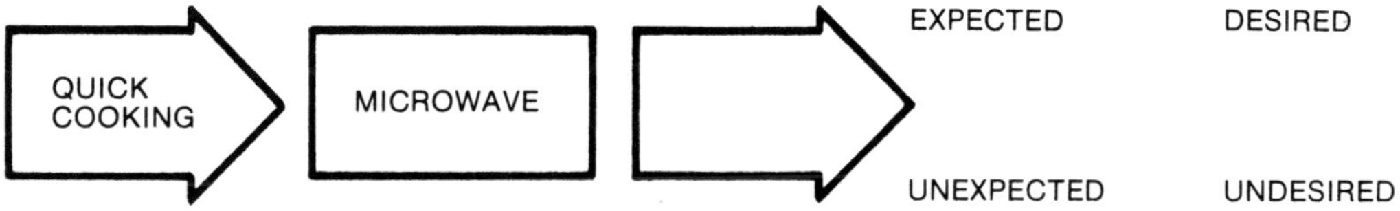

9. Respirator used in hospital to keep humans alive.

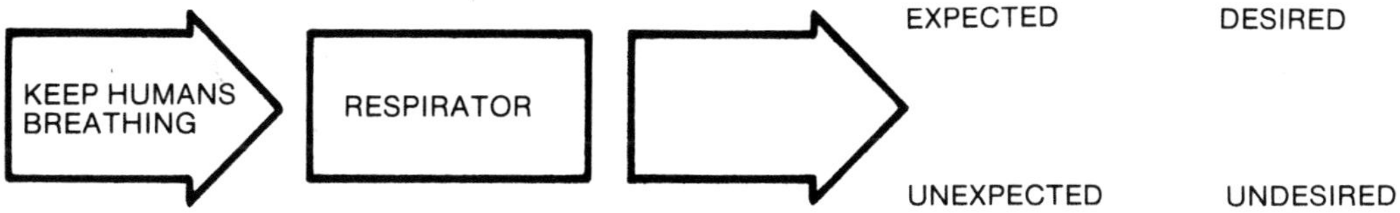

Name	Group Members
Selected Item	

Activity V-3
Technology and You

WHAT YOU ARE TO DO:

1. For one day, from the time you wake, you are to list all the technological inventions that affect you.
2. After one day of log keeping, you are to select one device for each of the technological aspects (bio-, informational, physical).
3. Show each of these items by using a technology process. Display the three. With each, show how the item does the following by stating:
   - How it affects you.
   - How you could improve or adapt it.
   - How you would replace it if you no longer had the item.
   - Type of system it is.
   - What resource item is mainly processed from.
   - All possible outputs for the item.

Use the next page for your log and aspect selection.

Name	Group Members
Selected Item	

Technology Effects Log

Morning

Noon

Evening

Items selected for each aspect.

Biotechnology ______________________________

Physical technology ______________________________

Informational technology ______________________________

Name	Group Members
Selected Item	

Activity V-4
City of the Future

WHAT YOU ARE TO DO:

In this activity, your class will design and build a City of the Future. In order to do this, you will follow the following steps:

1. The class will act as a town council.
2. The class will select a city mayor and a mayor's aide.
3. With the mayor and the mayor's aide leading the class through brainstorming, the city's needs will be stated. (For example, city needs energy, food, water.)
4. With all the city's needs and wants listed, the mayor and instructor will divide the class to two or three students assigned to one of the needs (group assigned to energy for the city).
5. Each group must create a system diagram of your solution to your group's assigned need, showing all outputs.
6. Each group will also set their device in the location assigned to them. This will show what the whole city looks like.

Needs and Wants Brainstorming

Need and want assigned to: ______________________________

Group members ______________________________

____

Scale size selected by city: ______________________________

Group's brainstorming on assigned need and want.

City of the Future—Need or Want

City of the Future—Need or Want

Name	Group Members
Selected Item	

Activity V-5 Future Living Centers

In this activity, your instructor will break the class into four groups. Each group will be assigned one of the following tasks:

1. Design and build a city of the future on land
2. Design and build a city of the future under water
3. Design and build a city of future in space
4. Design and build a transportation and communications system from one city to the other

When your group and task have been determined, your group is to design and build a solution to the task. Your solution must be stated in the system diagram and all outputs stated. Your group must explain how they adapted technology to meet their needs.

Name	Group Members
Selected Item	

Activity V-6
Hydroponic Greenhouse

WHAT YOU ARE TO DO:

1. Select the design you wish to use. Design 1, Design 2, or design your own.
2. Complete your design, add all dimensions to your design.
3. Layout and cut your material.
4. Assemble your hydroponic structure.
5. Design and build the systems you desire to have in your greenhouse.

Hydroponic Greenhouse Design 1

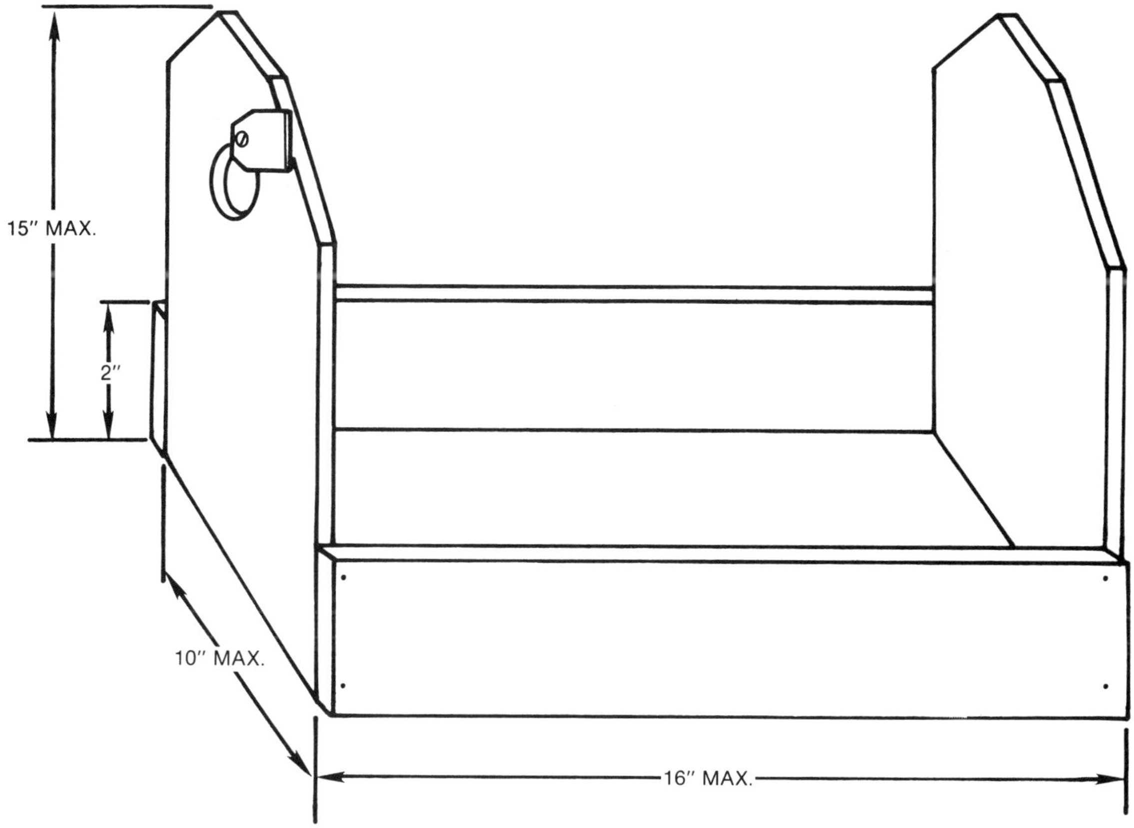

Hydroponic Greenhouse Design 2

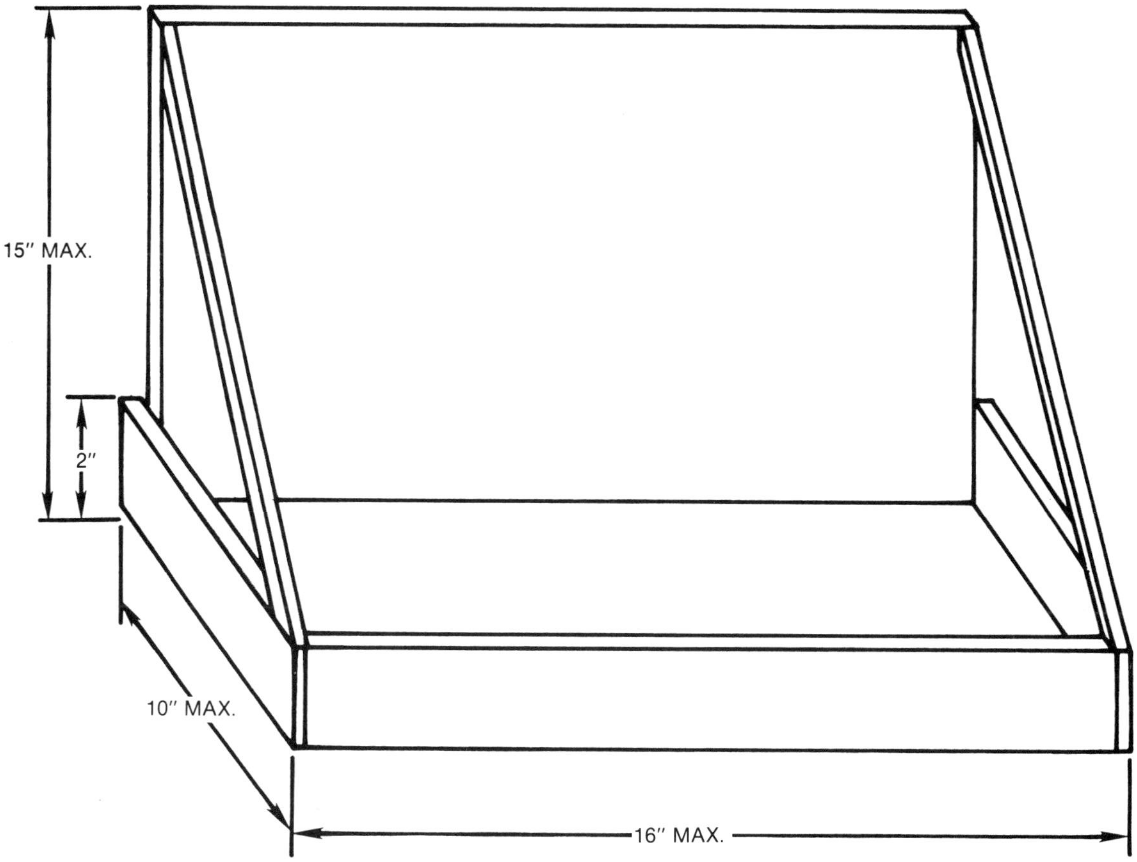

Devices to Add to the Hydroponic Greenhouse

1. Vacuum form tray or design and build a sheet metal tray.

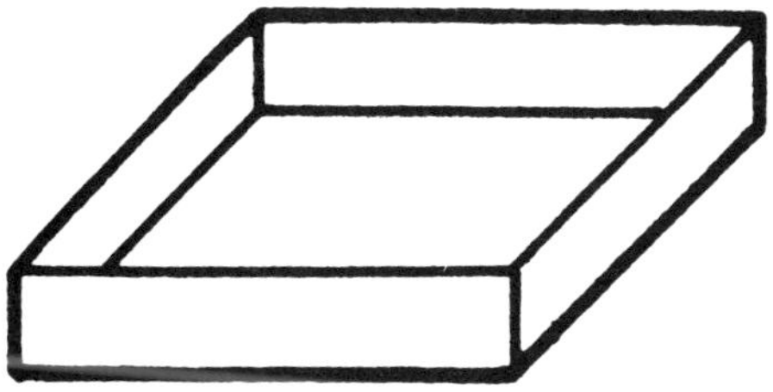

2. Add your own system to control temperature, moisture, and light.

EXAMPLE—Temperature Warning System

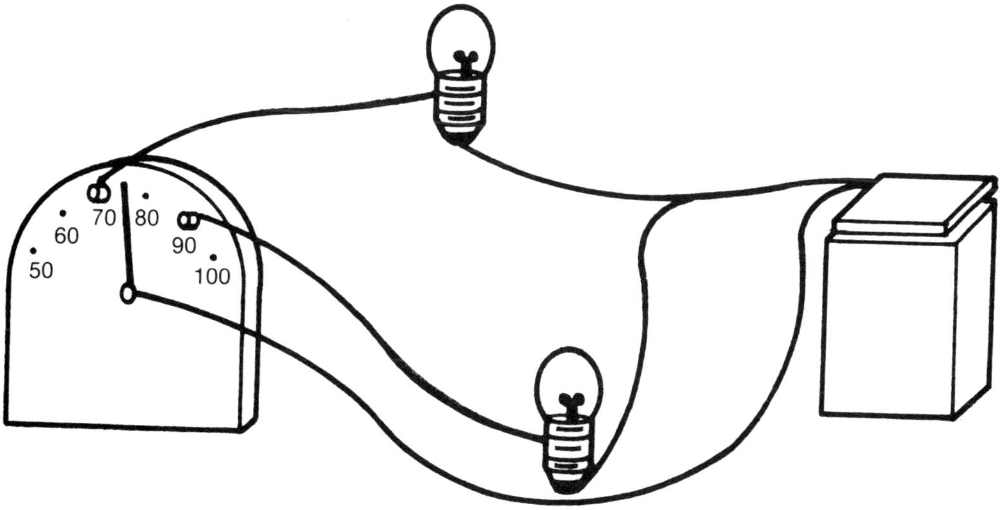

Note: You may replace lights with alarms. You may also replace lights with a heating or a cooling system.

APPENDIX

TOOLS AND MATERIALS PRICES

Below are the rental fees for tools, machines, materials, and other resources needed to solve your problem. These prices should be placed on a computer. Remember the parameters; some of the items below may not be used, depending upon the problem and its parameters.

OUTSIDE MATERIALS

Balloons $30.00 ea.
Cardstock $25.00 ea.
Cardstock Strips $5.00 ea.
Clear Plastic, 8″ × 10″ $1.00
Construction Paper $1.00 ea.
Copper Pipe $3.00/inch
Copper Wire $42.00/foot
DC Motor $15.00 ea.
Dowel, 1/4″–1/2″, 6″ long $1.00/2
Electrical Wire $1.00/foot
Eye Screws $5.00 ea.
LED $10.00 ea.
Light Bulbs $2.00 ea.
Light Sockets $10.00 ea.
Mouse Traps $30.00 ea.
Paper (not drafting) $1.00/5 sheets
Paper Cups $3.00 ea.
Pipe Cleaners $1.00 ea.
Popsicle Sticks 4 for $4.00
Rubberbands $5.00 ea.
Sheet Metal, 3″ × 8″ $5.00
Sheet Metal, 5″ × 8″ $10.00
Solder $1.00/inch
Straws 2 for $1.00
String $1.00/foot
Threaded Lamp Pipe $5.00/inch
Tongue Depressors $1.00/5
Wood, most types $10.00/inch
1/4″ MASONITE 6″ × 6″ $2.00
1/4″ × 2″ × 5″ Plywood $5.00 ea.
1/8″ MASONITE, 6′ × 6′ $5.00
8″ × 10″ Thermoplastic $5.00 ea.
9-Volt Battery Clip $5.00

LAYOUT/MEASUREMENT

Drafting Tools Free
Combination Square $1.00/10 min.
Erasers $5.00/10 min.
Framing Square $1.00/10 min.
Markers $1.00/color/20 min.
Scribe $1.00/10 min.
Sheet Metal Die $1.00/10 min.
Tri-Square $1.00/10 min.
Writing Devices $30.00

CUTTING AND MACHINING

All Hand Saws $1.00/cut or $5.00/10 min.
Aviation Snips $1.00/10 min.
Belt Sander $5.00/5 min.
Chisels $3.00/15 min.
Drill Press $5.00/hole
Files $1.00/20 min.
Hand Drill $1.00/hole
Hemmer $1.00/bend
Paper Cutter $1.00/4 cuts
$5.00/15 min.
Paper Cutter $5.00/10 min.
Power Hand Drill $3.00/hole
$5.00/10 min.
Power Hand Saws $3.00/cut
Power Hand Saws $5.00/15 min.
Razor Knife $5.00/10 min.
Sandpaper $1.00/sheet
Saw, Band $1.00/cut
Saws, Jig $5.00/10 min.
Scissors $1.00/10 min.
Sheet Metal Shears $5.00/10 min.
Tin Snips $1.00/10 min.

FASTENING

Brazing $10.00/25 min.
Brazing Rod $1.00/10 min.
Clamps, All Types $2.00/class
Dry Mount Press $5.00/print
Epoxy $10.00/5 min.
Glue Gun $5.00/10 min.
Glue Sticks $2.00 ea.
Hammer $1.00/10 min.
Mechanical Fasteners Screws $1.00 ea.
Mechanical Fasteners Bolts and Nuts $1.00 ea.
Nails, All Sizes $1.00/10
Paper Clips $1.00 ea.
Paper Fasteners $2.00/5
Rivet Gun $5.00/20 min.
Rivets, Any Size $1.00/5
Rubber Cement $1.00/10 min.
Sewing Methods $1.00/10 min.
Solder for Soldering $1.00/5″
Solder Gun $1.00/20 min.
Solder Pencil $1.00/10 min.
Soldering Torch $10.00/20 min.
Spot Welder $1.00/weld
Stapler Hand $10.00/10 staples
Stapler Paper $1.00/5 staples
Tape, Clear $5.00/6″
Tape, Electrical $5.00/6″
Tape, Masking $2.00/6″
White Glue $5.00/20 min.

FORMING AND MOLDING

Box and Pan $1.00/bend
Ceramic Materials "See Instructor"
Clear Cast Resin $10.00/mold
Hemmer, Hand $1.00/bend
Injection Molding $5.00/mold
Vacuum Former 1 mold—$5.00

COATING AND FINISHING

Deft, All $10.00/class
Markers See Layout/Measurement
Paint $5.00/10 min.
Paint Spray $15.00/5 min.
Stain, All $10.00/class
Varnish, All $10.00/class

MISCELLANEOUS ITEMS*

Bonus $5.00/great group work
Consultant Fee $10.00/10 min.
Credit Card $1.00/$25.00/class
Fine $5.00 for no safety glasses
Fine $10.00 for poor behavior
Patent Fee $10.00/application
Restrooms $1.00/visit
Savings $1.00/$50.00/class

*Consultant fee (outside information other than on tools and equipment) $10.00/15 min.